METHODS OF MICROARRAY
DATA ANALYSIS V

METHODS OF MICROARRAY DATA ANALYSIS V

Edited by

Patrick McConnell

Duke University

Simon M. Lin

Northwestern University

Patrick Hurban

Icoria, Inc.

 Springer

ISBN-13: 978-1-4419-4179-4
e-ISBN-13: 978-0-387-34569-7

Printed in acid-free paper.

9 8 7 6 5 4 3 2 1

springer.com

Contents

Contributors

I. AZZINI
Bioinformatics Group
SRA Division
ITC-irst
Via Sommarive 18
I-38050 Povo (TN), Italy

NICOLE E. BALDWIN
Department of Computer Science
The University of Tennessee
Knoxville, TN 37996

JUNIOR BARRERA
Instituto de Matemáticas
e Estatística e BIOINFO-UPS
Núcleo de Pesquisas em Bioinformática
Universidade de São Paulo
R. do Matão 1010
São Paulo, SP 05508-900, Brazil

E. BLANZIERI
Department of Information
and Communication Technology
University of Trento
Via Sommarive 14
I-3850 Povo (TN), Italy

JOSEPH BEYENE
Hospital for Sick
Children Research Institute
Department of Public Health Sciences
University of Toronto
555 University Avenue
Toronto, ON, M5G 1X8 Canada

ROBERTO M. CESAR, JR.
Instituto de Matemáticas
e Estatística e BIOINFO-UPS
Núcleo de Pesquisas em Bioinformática
Universidade de São Paulo
R. do Matão 1010
São Paulo, SP 05508-900, Brazil

LIANG CHEN
Department of Molecular,
Cellular, Developmental Biology
Yale University
New Haven, CT 06520

J.B. CHRISTIAN
Department of Statistics
Rice University
Houston, TX 77005

F. CIOCCHETTA
Bioinformatics Group
SRA Divison
ITC-irst and Department of ICT
University of Trento
Via Sommarive, 18
I-38050 Povo (TN), Italy

R. DELL'ANNA
Bioinformatics Group
SRA Division
ITC-irst
Via Sommarive 18
I-38050 Povo (TN), Italy

F. DEMICHELIS
Bioinformatics Group
SRA Divison
ITC-irst and Department of ICT
University of Trento
Via Sommarive, 18
I-38050 Povo (TN), Italy

CYR EMILE M'LAN
Department of Statistics
University of Connecticut
Storrs-Mansfield, CT 06269

N. GARG
Department of Microbiology
University of Texas Medical Branch
Galveston, TX 77555

CELIA M.T. GREENWOOD
Hospital for Sick
Children Research Institute
Department of Public Health Sciences
University of Toronto
555 University Avenue
Toronto, ON, M5G 1X8 Canada

R. GUERRA
Department of Statistics
Rice University
Houston, TX 77005

M.C. GUSTIN
Department of Biochemistry
and Cell Biology
Rice University
Houston, TX 77005

G. FOX
Department of Statistics
Rice University
Houston, TX 77005

PINGZHAO HU
Hospital for Sick
Children Research Institute
University of Toronto
555 University Avenue
Toronto, ON M5G 1X8 Canada

RAPHAEL D. ISOKPEHI, PH.D.
Department of Biology
Jackson State University
Jackson, MS 39217

RAYA KHANIN
Department of Statistics
University of Glasgow
Glasgow, UK

ANDREW V. KOSSENKOV
Bioinformatics Working Group
Fox Chase Cancer Center
Philadelphia, PA 19111

MICHAEL A. LANGSTON
Department of Computer Science
The University of Tennessee
Knoxville, TN 37996

FLORENCIA G. LEONARDI
Instituto de Matemáticas
e Estatística e BIOINFO-UPS
Núcleo de Pesquisas em Bioinformática
Universidade de São Paulo
R. do Matão 1010
São Paulo, SP 05508-900, Brazil

YIN LIU
Program of Computational Biology
and Bioinformatics
Yale University
New Haven, CT 06520

JUNFENG LIU
Department of Epidemiology
and Public Health
Yale University
New Haven, CT 06520

A. MALOSSINI
Department of Information
and Communication Technology
University of Trento
Via Sommarive 14
I-3850 Povo (TN), Italy

DAVID C. MARTINS, JR.
Instituto de Matemáticas
e Estatística e BIOINFO-UPS
Núcleo de Pesquisas em Bioinformática
Universidade de São Paulo
R. do Matão 1010
São Paulo, SP 05508-900, Brazil

MICHAEL McINTOSH
Department of Internal Medicine
Yale University
New Haven, CT 06520

EMILIO F. MERINO
Instituto de Ciências Biomedicas
Departamento de Parasitologia
Universidade de São Paulo
Ave. Lineu Prestes 1374
São Paulo, SP 05508-900, Brazil

GIRI NARASIMHAN
Bioinformatics Research Group
School of Computer Science
Florida International University
Miami, FL 33199

J. NOYOLA-MARTINEZ
Department of Statistics
Rice University
Houston, TX 77005

KALAI MATHEE
Department of Biological Sciences
School of Computer Science
Florida International University
Miami, FL 33199

MICHAEL F. OCHS
Bioinformatics Working Group
Fox Chase Cancer Center
Philadelphia, PA 19111

XINXIA PENG
Graduate School of Genome Science
and Technology
The University of Tennessee
Oak Ridge National Laboratory
Oak Ridge, TN 37831

CARLOS A. DE B. PEREIRA
Instituto de Matemáticas
e Estatística e BIOINFO-UPS
Núcleo de Pesquisas em Bioinformática
Universidade de São Paulo
R. do Matão 1010
São Paulo, SP 05508-900, Brazil

AIDAN J. PETERSON
Bioinformatics Working Group
Fox Chase Cancer Center
Philadelphia, PA 19111

HERNANDO A. DEL PORTILLO
Instituto de Ciências Biomedicas
Departamento de Parasitologia
Universidade de São Paulo
Ave. Lineu Prestes 1374
São Paulo, SP 05508-900, Brazil

A. ROMANEL
Department of Information
and Communication Technology
University of Trento
Via Sommarive 14
I-3850 Povo (TN), Italy

C. SHAW
Department of Molecular
and Human Genetics
Baylor College of Medicine
Houston, TX 77030

M. STEVENS
Department of Statistics
Department of Biochemistry
and Cell Biology
Rice University
Houston, TX 77005

ARNOLD M. SAXTON
Department of Animal Science
The University of Tennessee
Knoxville, TN 37996

A. SBONER
Bioinformatics Group
SRA Divison
ITC-irst and Department of ICT
University of Trento
Via Sommarive, 18
I-38050 Povo (TN), Italy

D.W. SCOTT
Department of Statistics
Rice University
Houston, TX 77005

JAY R. SNODDY
Graduate School of Genome Science
and Technology
The University of Tennessee
Oak Ridge National Laboratory
Oak Ridge, TN 37831

NING SUN
Department of Epidemiology
and Public Health
Yale University
New Haven, CT 06520

HONGYU ZHAO
Department of Epidemiology
and Public Health
Department of Genetics
Yale University
New Haven, CT 06520

LIANGBIAO ZHENG
Department of Epidemiology
and Public Health
Yale University
New Haven, CT 06520

ERLIANG ZENG
Bioinformatics Research Group
School of Computer Science
Florida International University
Miami, FL 33199

RICARDO Z.N. VENCIO
Instituto de Matemáticas
e Estatística e BIOINFO-UPS
Núcleo de Pesquisas em Bioinformática
Universidade de São Paulo
R. do Matão 1010
São Paulo, SP 05508-900, Brazil

ERNEST WIT
Department of Statistics
University of Glasgow
Glasgow, UK

CHENGYONG YANG
Bioinformatics Research Group
School of Computer Science
Florida International University
Miami, FL 33199

MÁRCIO M. YAMAMOTO
Instituto de Ciências Biomedicas
Departamento de Parasitologia
Universidade de São Paulo
Ave. Lineu Prestes 1374
São Paulo, SP 05508-900, Brazil

INTRODUCTION

As a technology platform, microarrays have been transformed into a fundamental component of modern biological research. Laboratory methods have continually improved, but many of the most important advances in the field have come from the development of enhanced analysis methods. The evolution of microarray data analysis is brought into sharp focus on an annual basis by the Critical Assessment of Microarray Data Analysis (CAMDA) Conference, with the fifth edition taking place during November of 2004. Like previous gatherings, CAMDA 2004 featured a melange of disciplines – biologists, statisticians, bioinformaticians, computer scientists and mathematicians were all in attendance – and had an international flavor, with researchers from 11 countries joining in the discussion.

The subject of this year's contest dataset was the intraerythrocytic developmental cycle of *Plasmodium falciparum*, the most deadly of malarial parasites, and was kindly provided by the DeRisi lab at UCSF. Malaria infections number 300–500 million every year, and result in 1–2 million deaths, 90% of which occur in sub-Saharan Africa. It is estimated that upwards of 40% of the world's population is threatened by malaria, and that malaria is responsible for perhaps 20% of childhood deaths before the age of 5 in Africa. Thus, the potential impact of research into the life cycle of *Plasmodium* cannot be overstated.

CAMDA participants were treated to an excellent keynote address from Dr. Manuel Llinás, an author of the originating work. His address set the stage for a lively debate on the merits of 16 papers presenting innovative methods for the analysis of these data. Importantly, the emphasis this year was on the application of these methods towards the elucidation of biological questions. In the end, attendees split the vote for best presentation between two equally deserving groups:

J. Barrera, R.M. Cesar Jr., D.C. Martins Jr., E.F. Merino, R.Z.N. Vencio, F.G. Leonardi, M.M. Yamamoto, C.A.B. Pereira, and H.A. del Portillo, University of Sao Paulo, Sao Paulo, Brazil, "A New Annotation Tool for Malaria Based on Inference of Probabilistic Genetic Networks"

and

J.B. Christian, C. Shaw, J. Noyola-Martinez, M.C. Gustin, D.W. Scott, and R. Guerra, Rice University and Baylor College of Medicine, "Spatial Correlation of Expression in *P. falciparum*"

In addition to these presentations, The CAMDA Scientific Committee has compiled a number of outstanding papers into this volume. We hope that you find the insights presented useful, and that you join us for the next CAMDA Conference, which is itself evolving from the analysis of only microarray data into a forum for the integrative analysis of multiple data streams.

Patrick McConnell
Simon M. Lin
Patrick Hurban

ACKNOWLEDGEMENTS

The editors thank the contributing authors for their very fine work. We also acknowledge Emily Allred for her tireless work and dedication to organizing the CAMDA conference. We thank our supporters at Duke University, especially the Duke Comprehensive Cancer Center. The CAMDA conference and these proceedings would not be possible without the contributions of the scientific committee and other reviewers (listed below) who contribute to the scientific review process. Our heartfelt thanks for the time and effort they commit to CAMDA. We especially thank our corporate sponsors for their generous support: Agilent Technologies and The GenomeWeb Intelligence Network (bio1nf0rm and bioArray News).

Reviewers

Andrew Allen
Bruce Aronow
Philippe Broet
Yong Chen
Kevin Coombes
Christopher Corton
David Dix
Joaquin Dopazo
Werner Dubitzky
Greg Gibson
J. Gormley
Greg Grant
Patrick Hurban
Elena Kleymenova
David Kreil
Kwan Lee

Simon M. Lin
Delong Liu
James Lyons-Weiler
Steve Marron
Patrick McConnell
Geoff McLachlan
Kouros Owzar
Joel Parker
Aidan Peterson
Raymond Samaha
Imran Shah
Jennifer Shoemaker
Rainer Spang
Paul Spellman
Thomas Wu

Chapter 1

Data Mining of Malaria Parasite Gene Expression for Possible Translational Research

Raphael D. Isokpehi

Department of Biology, Jackson State University, Jackson, MS 39217, USA

Abstract   Malaria is a caused by protozoan parasites of the genus *Plasmodium* where the host is a vertebrate and the vector is a mosquito. Over 1 million deaths are attributed to malaria each year. Drug-resistant parasites, insecticide-resistant mosquitoes, and the lack of an effective vaccine to protect the host are major impediments to the prevention and control of malaria. Genome-wide microarray experiments on malaria parasites have provided new insights into the transcription of genes and gene networks during parasite development stages. This overview article presents basic information on malaria parasites, their hosts and vectors; life cycle of human malaria parasites; and gene content of malaria parasites. Mathematical and statistical methods presented at conferences on Critical Assessment of Microarray Data Analysis (CAMDA) could be integrated into malaria parasite databases as tools for widespread use thereby accelerating the characterization of biological processes that are basis for better drugs, effective vaccines and easy-to-use diagnostics.

1.   INTRODUCTION

Malaria is a mosquito-transmitted parasitic disease of major global public health concern. Annually, there are at least 300 million acute illnesses and over 1 million deaths from the disease (Breman et al., 2004; Snow et al., 2005). Malaria kills a child every 30 seconds (Webster, 2001). Transmission of malaria occurs in at least 107 countries and territories with 3.2 billion people at risk (WHO, 2005). Countries at risk are in tropical regions of Africa, the Americas and Asia. Transmission is highest in sub-Saharan Africa because of the warm climate that encourages the survival of the mosquitoes that carry the parasites. Children under the age of 5 years and pregnant women are the populations at highest risk of malaria because of their inability to mount an effective immune defense response to the parasite infection. Travelers from malaria-free countries to disease-endemic countries are also at risk of contracting malaria because of naïve immunity to malaria infection.

In the past 7 years substantial progress has been made by global private and governmental initiatives to control and prevent malaria (Narasimhan and Attaran, 2003; Nwaka, 2005; Olliaro, 2005; Olumese, 2005; WHO, 2005). However, there are still challenges in making further progress such as parasite resistance to inexpensive malaria drugs; lack of an effective vaccine; the need for inexpensive easy-to-use diagnostics that will prevent misuse of malaria drugs and an effective spraying method in the environment to prevent mosquito larva from hatching. From a basic research perspective, the availability of the genome sequences of malaria parasites, their hosts and vectors (Carlton et al., 2002; Gardner et al., 2002; Hall et al., 2005; Hoffman et al., 2002; Mongin et al., 2004) as well as measurements of gene and protein expression levels of parasite developmental stages (Bozdech et al., 2003; Daily et al., 2004; Florens et al., 2002; Hall et al., 2005; Khan et al., 2005; Lasonder et al., 2002; Le Roch et al., 2004, 2003; Silvestrini et al., 2005; Young et al., 2005) provides unprecedented research resources to identify biological process that could be targeted for development of new drugs, effective vaccines, diagnostics and treatment strategies.

The Fifth International Conference for the Critical Assessment of Microarray Data Analysis (CAMDA 2004; http://www.camda.duke.edu/camda04) featured presentations on sophisticated mathematical and statistical methods encoded in computer programs that were applied to microarray gene profiling datasets on the 48-hour asexual blood stage development of a malaria parasite *Plasmodium falciparum* (Llinas and del Portillo, 2005). The articles from CAMDA 2004 reported in this volume of Methods in Microarray Data Analysis represent a major contribution to data analysis of the ever growing datasets from studies on malaria parasite genes. This overview article aims to complement the conference articles by presenting basic information on (i) malaria parasites, their hosts and vectors; (ii) life cycle of human malaria parasites; and (iii) gene content of malaria parasites. The overview concludes on how the computational methods from CAMDA 2004 may find widespread use in malaria research.

2.   MALARIA PARASITES, HOSTS AND VECTORS

Parasitic protozoa of the genus *Plasmodium* (*P.*) cause malaria. Malaria parasites can grow and develop within host cells with a complex life cycle involving multiple stages in an invertebrate insect vector and a vertebrate host (Baton and Ranford-Cartwright, 2005). Vertebrate hosts of malaria parasites include humans, non-human primates, rodents, birds and reptiles (Table 1).

Table 1. Malaria parasites and hosts

Human	Non-human Primate	Rodent	Avian
Plasmodium falciparum	*Plasmodium cynomolgi*	*Plasmodium berghei*	*Plasmodium gallinaceum*
Plasmodium malariae	*Plasmodium fieldi*	*Plasmodium chabaudi*	*Plasmodium elongatum*
Plasmodium ovale	*Plasmodium inui*	*Plasmodium yoelii*	
Plasmodium vivax	*Plasmodium knowlesi*		
	Plasmodium reichenowi		
	Plasmodium simiovale		
	Plasmodium simium		

Table 2. Geographical distribution of selected *Anopheles* mosquitoes known to transmit human malaria parasites

Africa	Asia	Pacific area	Americas
Anopheles gambiae	*Anopheles culicifaciens*	*Anopheles farauti*	*Anopheles albimanus*
Anopheles funestus	*Anopheles dirus*	*Anopheles maculatus*	*Anopheles darlingi*
	Anopheles sinensis		
	Anopheles miminus		

Four species *Plasmodium falciparum*, *Plasmodium vivax*, *Plasmodium ovale* and *Plasmodium malariae* are known to naturally infect humans. Two human parasites *P. falciparum* and *P. vivax* are the two most commonly encountered. Both species can invade liver and red blood cells resulting in clinical manifestations that can range from asymptomatic to death depending on a combination of host, parasite, geographical and social factors (Miller et al., 2002). The most deadly of the human malaria parasites is *Plasmodium falciparum* being responsible for 90% of malaria cases and deaths in Africa. Human malaria parasites are able to evade the host immune response by diverse mechanisms including antigen polymorphism, antigenic variation and immune modulation (Ferreira et al., 2004; Hisaeda et al., 2005).

Non-human malaria parasites such as *P. berghei*, *P. chabaudi*, *P. knowlesi* and *P. yoelii* have been useful as models for studying malaria infection in humans (Carlton and Carucci, 2002). The rodent malaria parasites have been particularly useful for studying parasite stages in the mosquito and liver as well as drug resistance (Cravo et al., 2003).

Female mosquitoes of the genus *Anopheles* are the principal vector of human malaria parasites (Kiszewski et al., 2004). *Anopheles gambiae* which is major vector of *P. falciparum* in Africa is a complex of sibling species. Table 2 shows geographical distribution of selected *Anopheles* species known to transmit human malaria parasites.

3.   LIFE CYCLE OF HUMAN MALARIA PARASITES

The major components of a parasite's life cycle are growth, development, transmission and reproduction. The life cycle of malaria parasites is complicated because of multiple stages and varying duration in the host and the vector (Baton and Ranford-Cartwright, 2005; Kappe et al., 2004; Talman et al., 2004). There are three critical invasive stages in the life cycle of human malaria parasites that lead to parasite multiplication and destruction of host cells. These invaded cells are the hepatocytes (exoerythrocytic or liver stage) and erythrocytes (intraerythrocytic or blood stage), and the mosquito midgut epithelium (sporogonic stage). Parasite forms in the life cycle could either be produced by asexual replication or sexual reproduction. The exoerythrocytic development involves only asexual replication while the intraerythrocytic development involves asexual replication as well as the formation of male and female gamete precursors or gametocytes. The asexual intraerythrocytic stage is responsible for the clinical symptoms of the disease such as fever, weakness, pains and chills. Complications of infection by *P. falciparum* in humans include severe anemia and cerebral malaria. Gametocytes continue the cycle in the mosquito by maturing to male and female gametes. The mature gametes that are fertilized eventually develop to sporozoites that can invade liver cells.

4.   GENE CONTENT OF MALARIA PARASITES

The landmark publications in October 2002 of the A-T rich (~80%) genome sequences of *P. falciparum* and *P. yoelii* revealed new insights and new research resources to study the biology of malaria parasites (Carlton et al., 2002; Gardner et al., 2002). The genome sequence of *P. vivax* has been determined (Carlton, 2003). The available *P. vivax* genome data has already shown novel genes shared with *P. falciparum* that can be used to study immune responses to *P. vivax* during liver stages in malaria endemic settings (Wang et al., 2005).

The integration of data from the genomes of *P. falciparum* and *P. yoelii* with partial genome sequence information, gene microarrays and protein expression studies of two rodent malaria parasites *P. berghei* and *P. chabaudi* has enriched the understanding of stage-specific genes in the *Plasmodium* life cycle (Hall et al., 2005). Major observations from these studies on the gene content of malaria parasites are that their nuclear genome is about 23 megabases distributed over 14 chromosomes. Furthermore, the total number of genes is over 5,000 with about 4,500 genes shared by *P. falciparum, P. yoelii, P. berghei* and *P. chabaudi* (Carlton et al., 2002; Hall et al., 2005). Thus, data collection from studies employing non-human malaria parasites may be extrapolated to malaria infection in humans.

As of the time of publishing the genome of *P. falciparum*, about 60% of the parasite's genes could not be assigned a function. The proportion of genes of unknown function is likely to have reduced because of the availability of additional genomes of malaria parasites, high-throughput gene and protein expression assays as well as powerful methods for detecting orthologous and paralogous genes. These methodological approaches have improved the annotation of *P. falciparum* genome (Li et al., 2003b; Sam-Yellowe et al., 2004; Yeh et al., 2004). For example, Yeh et al. (2004) used computational methods to improve the annotation of 956 hypothetical proteins. Furthermore, Sam-Yellowe et al. (2004) used shared stretches of amino acid sequences from proteomics experiments to manually re-annotate 10 genes (PFA0680c, PFA0065w, PFB0985c, PFC1080c, MAL6P1.15, MAL7P1.5, MAL7P1.58, PF10_0390, PF11_0014 and PF11_0025) as members of a gene family. The gene list obtained by Sam-Yellowe et al. (2004) also agrees with the predicted orthologous–paralogous grouping of Li et al. (2003b). All these reports illustrate how multiple approaches can provide evidence for gene function in malaria parasites.

5. INTEGRATION OF METHODS FROM CAMDA CONFERENCES INTO MALARIA PARASITE DATABASES

Knowledge about stage and time a parasite gene is expressed could aid in identifying novel biological processes that are candidate drug, diagnostic and antigenic targets. The generation of Expressed Sequence Tags (ESTs) and Genome Survey Sequences (GSS) from cDNA libraries derived from parasite stages allowed the survey of gene expression as well as initial microarray-based gene expression profiling (Ben et al., 2001; Carlton et al., 2001; Hayward et al., 2000; Li et al., 2003a). Furthermore, the availability of genome sequences of malaria parasites has facilitated the genome-wide measurement of mRNA levels during the life cycle stages (Bozdech et al., 2003; Daily et al., 2004; Florens et al., 2002; Hall et al., 2005; Khan et al., 2005; Lasonder et al., 2002; Le Roch et al., 2004, 2003; Silvestrini et al., 2005; Young et al., 2005). Thus, various methodological breakthroughs have allowed for new datasets from parasite stages which previously where difficult study.

Powerful computational methods are vital to predicting the biological significance of expression profiles observed from gene expression experiments. The selection of microarray experiments of the Intraerythrocytic Development

Table 3. Rank Correlation Coefficients of gene
expression profiles for three strains of *Plasmodium
falciparum*[a]

	HB3	3D7	Dd2
HB3	1		
3D7	0.97	1	
Dd2	0.92	0.94	1

[a] Rank Correlation Coefficients of gene expression during intraerythrocytic development cycle was calculated using expression correlation
tool on PlasmoDB.

Cycle (IDC) (Bozdech et al., 2003) for data analysis by the organizers of the
Fifth International Conference for the Critical Assessment of Microarray Data
Analysis (CAMDA 2004) has allowed for a variety of mathematical and statistical techniques to be applied to a dataset on the 48-hour asexual development of *P. falciparum* in red blood cells. Novel insights into the biology
of *P. falciparum* reported in the conference articles include (i) how periodically expressed genes are distributed on chromosomes; (ii) potential transcription factor binding motifs; (iii) additional genes with periodic expression; (iv)
merozoite invasion genes showing anomalous peaks of expression during the
time course; (v) gene hubs of essential or lethal genes in gene networks; (vi)
regulatory system of glycolysis; (vii) co-expressed genes with same combination of protein domains; and (vii) sequence motifs that may regulate oxidative
stress genes.

The complex life cycle of malaria parasites, drug resistance, lack of an effective vaccine and insecticide resistance necessitates the discovery of additional
biological insights. The methods described in the CAMDA 2004 conference
articles could be integrated as tools in database resources that provide access
to genome and genome-derived datasets on malaria parasites.

To illustrate the usefulness of such integration, a researcher could ask: What
is the correlation of gene expression profiles for a predicted gene across parasites strains? To answer this question, the researcher could use the expression
profile correlation tool on the *Plasmodium* Genome Resource (PlasmoDB;
http://plasmodb.org/) to obtain Rank Correlation Coefficients (RCC) for the
expression of the gene during intraerythrocytic development across three parasite strains HB3, 3D7 and Dd2 (Bozdech et al., 2003). The strains originate
from different regions of the world and have known drug resistance to chloro-
quine, sulfadoxine and pyrimethanine. Information that a gene is expressed in
a similar manner in all strains may support the essentiality of gene for parasite
survival.

As an example to elucidate on the need to integrate CAMDA methods into malaria databases, the expression profile correlation across the strains for the nuclear gene PF07_0087 was determined to further characterize the gene as a possible drug target. The protein product of PF07_0087 is targeted to the apicoplast (Foth et al., 2003) and is differentially expressed in asexual blood stages compared to gametocyte blood stages (Isokpehi and Hide, 2003). The apicoplast is an essential organelle present in malaria parasites and other apicomplexan parasites such as *Cryptosporidium, Eimeria* and *Toxoplasma* (Waller and McFadden, 2005). The metabolic pathways in the apicoplasts are targets for new antimalarial drug development (Sato and Wilson, 2005). The RCC for the gene from pairwise comparison of strains is close to 1.00 (Table 3) indicating similar gene expression profiles of PF07_0087 during IDC for the three strains. Since the calculation of the RCC statistic is available in PlasmoDB as a tool, the results can be integrated with those from other tools thereby enhancing the understanding of the gene or sets of genes being investigated.

ACKNOWLEDGEMENTS

The author thanks Hari Cohly for useful comments and acknowledges support from grant G12RR13459 from the National Institutes of Health to the Center for Environmental Health, Jackson State University.

REFERENCES

Baton, L.A. and Ranford-Cartwright, L.C. (2005), Spreading the seeds of million-murdering death: metamorphoses of malaria in the mosquito, *Trends Parasitol.*

Ben, M.C., Gluzman, I.Y., Hott, C., MacMillan, S.K., Amarakone, A.S., Anderson, D.L., Carlton, J.M., Dame, J.B., Chakrabarti, D., Martin, R.K., Brownstein, B.H., and Goldberg, D.E. (2001), Co-ordinated programme of gene expression during asexual intraerythrocytic development of the human malaria parasite *Plasmodium falciparum* revealed by microarray analysis, *Mol. Microbiol.*, **39**, 26–36.

Bozdech, Z., Llinas, M., Pulliam, B.L., Wong, E.D., Zhu, J., and Derisi, J.L. (2003), The transcriptome of the intraerythrocytic developmental cycle of *Plasmodium falciparum, PLoS Biol.*, **1**, E5.

Breman, J.G., Alilio, M.S., and Mills, A. (2004), Conquering the intolerable burden of malaria: What's new, what's needed: A summary, *Am. J. Trop. Med. Hyg.*, **71**, 1–15.

Carlton, J.M. and Carucci, D.J. (2002), Rodent models of malaria in the genomics era, *Trends Parasitol.*, **18**, 100–102.

Carlton, J.M., Muller, R., Yowell, C.A., Fluegge, M.R., Sturrock, K.A., Pritt, J.R., Vargas-Serrato, E., Galinski, M.R., Barnwell, J.W., Mulder, N., Kanapin, A., Cawley, S.E., Hide, W.A., and Dame, J.B. (2001), Profiling the malaria genome: A gene survey of three species of malaria parasite with comparison to other apicomplexan species, *Mol. Biochem. Parasitol.*, **118**, 201–210.

Carlton, J. (2003), The *Plasmodium vivax* genome sequencing project, *Trends Parasitol.*, **19**, 227–231.

Carlton, J.M., Angiuoli, S.V., Suh, B.B., Kooij, T.W., Pertea, M., Silva, J.C., Ermolaeva, M.D., Allen, J.E., Selengut, J.D., Koo, H.L., Peterson, J.D., Pop, M., Kosack, D.S., Shumway, M.F., Bidwell, S.L., Shallom, S.J., van Aken, S.E., Riedmuller, S.B., Feldblyum, T.V., Cho, J.K., Quackenbush, J., Sedegah, M., Shoaibi, A., Cummings, L.M., Florens, L., Yates, J.R., Raine, J.D., Sinden, R.E., Harris, M.A., Cunningham, D.A., Preiser, P.R., Bergman, L.W., Vaidya, A.B., van Lin, L.H., Janse, C.J., Waters, A.P., Smith, H.O., White, O.R., Salzberg, S.L., Venter, J.C., Fraser, C.M., Hoffman, S.L., Gardner, M.J., and Carucci, D.J. (2002), Genome sequence and comparative analysis of the model rodent malaria parasite *Plasmodium yoelii yoelii*, *Nature*, **419**, 512–519.

Cravo, P.V., Carlton, J.M., Hunt, P., Bisoni, L., Padua, R.A., and Walliker, D. (2003), Genetics of mefloquine resistance in the rodent malaria parasite *Plasmodium chabaudi*, *Antimicrob Agents Chemother.* **47**, 709–718.

Daily, J.P., Le Roch, K.G., Sarr, O., Fang, X., Zhou, Y., Ndir, O., Mboup, S., Sultan, A., Winzeler, E.A., and Wirth, D.F. (2004), In vivo transcriptional profiling of *Plasmodium falciparum*, *Malar J.* **3**, 30, 30.

Ferreira, M.U., da Silva, N.M., and Wunderlich, G. (2004), Antigenic diversity and immune evasion by malaria parasites, *Clin. Diagn. Lab. Immunol.*, **11**, 987–995.

Florens, L., Washburn, M.P., Raine, J.D., Anthony, R.M., Grainger, M., Haynes, J.D., Moch, J.K., Muster, N., Sacci, J.B., Tabb, D.L., Witney, A.A., Wolters, D., Wu, Y., Gardner, M.J., Holder, A.A., Sinden, R.E., Yates, J.R., and Carucci, D.J. (2002), A proteomic view of the *Plasmodium falciparum* life cycle, *Nature*, **419**, 520–526.

Foth, B.J., Ralph, S.A., Tonkin, C.J., Struck, N.S., Fraunholz, M., Roos, D.S., Cowman, A.F., and McFadden, G.I. (2003), Dissecting apicoplast targeting in the malaria parasite *Plasmodium falciparum*, *Science*, **299**, 705–708.

Gardner, M.J., Hall, N., Fung, E., White, O., Berriman, M., Hyman, R.W., Carlton, J.M., Pain, A., Nelson, K.E., Bowman, S., Paulsen, I.T., James, K., Eisen, J.A., Rutherford, K., Salzberg, S.L., Craig, A., Kyes, S., Chan, M.S., Nene, V., Shallom, S.J., Suh, B., Peterson, J., Angiuoli, S., Pertea, M., Allen, J., Selengut, J., Haft, D., Mather, M.W., Vaidya, A.B., Martin, D.M., Fairlamb, A.H., Fraunholz, M.J., Roos, D.S., Ralph, S.A., McFadden, G.I., Cummings, L.M., Subramanian, G.M., Mungall, C., Venter, J.C., Carucci, D.J., Hoffman, S.L., Newbold, C., Davis, R.W., Fraser, C.M., and Barrell, B. (2002), Genome sequence of the human malaria parasite *Plasmodium falciparum*, *Nature*, **419**, 498–511.

Hall, N., Karras, M., Raine, J.D., Carlton, J.M., Kooij, T.W., Berriman, M., Florens, L., Janssen, C.S., Pain, A., Christophides, G.K., James, K., Rutherford, K., Harris, B., Harris, D., Churcher, C., Quail, M.A., Ormond, D., Doggett, J., Trueman, H.E., Mendoza, J., Bidwell, S.L., Rajandream, M.A., Carucci, D.J., Yates III, J.R., Kafatos, F.C., Janse, C.J., Barrell, B., Turner, C.M., Waters, A.P., and Sinden, R.E. (2005), A comprehensive survey of the *Plasmodium* life cycle by genomic, transcriptomic, and proteomic analyses, *Science*, **307**, 82–86.

Hayward, R.E., Derisi, J.L., Alfadhli, S., Kaslow, D.C., Brown, P.O., and Rathod, P.K. (2000), Shotgun DNA microarrays and stage-specific gene expression in *Plasmodium falciparum* malaria, *Mol. Microbiol.*, **35**, 6–14.

Hisaeda, H., Yasutomo, K., and Himeno, K. (2005), Malaria: Immune evasion by parasites, *Int. J. Biochem. Cell Biol.*, **37**, 700–706.

Hoffman, S.L., Subramanian, G.M., Collins, F.H., and Venter, J.C. (2002), *Plasmodium*, human and *Anopheles* genomics and malaria, *Nature*, **415**, 702–709.

Isokpehi, R.D. and Hide, W.A. (2003), Integrative analysis of intraerythrocytic differentially expressed transcripts yields novel insights into the biology of *Plasmodium falciparum*, *Malar. J.*, **2**, 38.

Kappe, S.H., Buscaglia, C.A., and Nussenzweig, V. (2004), *Plasmodium* sporozoite molecular cell biology, *Annu. Rev. Cell. Dev. Biol.*, **20**, 29–59.

Khan, S.M., Franke-Fayard, B., Mair, G.R., Lasonder, E., Janse, C.J., Mann, M., and Waters, A.P. (2005), Proteome analysis of separated male and female gametocytes reveals novel sex-specific *Plasmodium* biology, *Cell*, **121**, 675–687.

Kiszewski, A., Mellinger, A., Spielman, A., Malaney, P., Sachs, S.E., and Sachs, J. (2004), A global index representing the stability of malaria transmission, *Am. J. Trop. Med. Hyg.*, **70**, 486–498.

Lasonder, E., Ishihama, Y., Andersen, J.S., Vermunt, A.M., Pain, A., Sauerwein, R.W., Eling, W.M., Hall, N., Waters, A.P., Stunnenberg, H.G., and Mann, M. (2002), Analysis of the *Plasmodium falciparum* proteome by high-accuracy mass spectrometry, *Nature*, **419**, 537–542.

Le Roch, K.G., Zhou, Y., Blair, P.L., Grainger, M., Moch, J.K., Haynes, J.D., De V, L., Holder, A.A., Batalov, S., Carucci, D.J., and Winzeler, E.A. (2003), Discovery of gene function by expression profiling of the malaria parasite life cycle, *Science*, **301**, 1503–1508.

Le Roch, K.G., Johnson, J.R., Florens, L., Zhou, Y., Santrosyan, A., Grainger, M., Yan, S.F., Williamson, K.C., Holder, A.A., Carucci, D.J., Yates III, J.R., and Winzeler, E.A. (2004), Global analysis of transcript and protein levels across the *Plasmodium falciparum* life cycle, *Genome Res.*, **14**, 2308–2318.

Li, L., Brunk, B.P., Kissinger, J.C., Pape, D., Tang, K., Cole, R.H., Martin, J., Wylie, T., Dante, M., Fogarty, S.J., Howe, D.K., Liberator, P., Diaz, C., Anderson, J., White, M., Jerome, M.E., Johnson, E.A., Radke, J.A., Stoeckert Jr., C.J., Waterston, R.H., Clifton, S.W., Roos, D.S., and Sibley, L.D. (2003a). Gene discovery in the apicomplexa as revealed by EST sequencing and assembly of a comparative gene database, *Genome Res.*, **13**, 443–454.

Li, L., Stoeckert Jr., C.J., and Roos, D.S. (2003b). OrthoMCL: Identification of ortholog groups for eukaryotic genomes, *Genome Res.*, **13**, 2178–2189.

Llinas, M., and del Portillo, H.A. (2005), Mining the malaria transcriptome, *Trends Parasitol.*, **21**, 350–352.

Miller, L.H., Baruch, D.I., Marsh, K., and Doumbo, O.K. (2002), The pathogenic basis of malaria, *Nature*, **415**, 673–679.

Mongin, E., Louis, C., Holt, R.A., Birney, E., and Collins, F.H. (2004), The *Anopheles* gambiae genome: An update, *Trends Parasitol.*, **20**, 49–52.

Narasimhan, V., and Attaran, A. (2003), Roll back malaria? The scarcity of international aid for malaria control, *Malar. J.*, **2**, 8.

Nwaka, S. (2005), Drug discovery and beyond: The role of public-private partnerships in improving access to new malaria medicines, *Trans. R. Soc. Trop. Med. Hyg.*, **99**(Suppl 1), S20–S29.

Olliaro, P. (2005), Drug resistance hampers our capacity to roll back malaria, *Clin. Infect. Dis.* **41**(Suppl 4), S247–S257.

Olumese, P. (2005), Epidemiology and surveillance: Changing the global picture of malaria – myth or reality? *Acta Trop.*, **95**, 265–269.

Sam-Yellowe, T.Y., Florens, L., Johnson, J.R., Wang, T., Drazba, J.A., Le Roch, K.G., Zhou, Y., Batalov, S., Carucci, D.J., Winzeler, E.A., and Yates III, J.R. (2004), A *Plasmodium* gene family encoding Maurer's cleft membrane proteins: Structural properties and expression profiling, *Genome Res.*, **14**, 1052–1059.

Sato, S., and Wilson, R.J. (2005), The plastid of *Plasmodium* spp.: A target for inhibitors, *Curr. Top. Microbiol. Immunol.*, **295**, 251–273.

Silvestrini, F., Bozdech, Z., Lanfrancotti, A., Di, G.E., Bultrini, E., Picci, L., Derisi, J.L., Pizzi, E., and Alano, P. (2005), Genome-wide identification of genes upregulated at the onset of gametocytogenesis in *Plasmodium falciparum*, *Mol. Biochem. Parasitol.*, **143**, 100–110.

Snow, R.W., Guerra, C.A., Noor, A.M., Myint, H.Y., and Hay, S.I. (2005), The global distribution of clinical episodes of *Plasmodium falciparum* malaria, *Nature*, **434**, 214–217.

Talman, A.M., Domarle, O., McKenzie, F.E., Ariey, F., and Robert, V. (2004), Gametocytogenesis: The puberty of *Plasmodium falciparum*, *Malar. J.* **3** (24), 24.

Waller, R.F., and McFadden, G.I. (2005), The apicoplast: A review of the derived plastid of apicomplexan parasites, *Curr. Issues Mol. Biol.*, **7**, 57–79.

Wang, R., Revalo-Herrera, M., Gardner, M.J., Bonelo, A., Carlton, J.M., Gomez, A., Vera, O., Soto, L., Vergara, J., Bidwell, S.L., Domingo, A., Fraser, C.M., and Herrera, S. (2005), Immune responses to *Plasmodium vivax* pre-erythrocytic stage antigens in naturally exposed Duffy-negative humans: A potential model for identification of liver-stage antigens, *Eur. J. Immunol.*, **35**, 1859–1868.

Webster, D. (2001), Malaria kills one child every 30 seconds, *J. Public. Health Policy*, **22**, 23–33.

WHO (2005), World Malaria Report 2005, World Health Organization, Geneva.

Yeh, I., Hanekamp, T., Tsoka, S., Karp, P.D., and Altman, R.B. (2004), Computational analysis of *Plasmodium falciparum* metabolism: Organizing genomic information to facilitate drug discovery, *Genome Res.*, **14**, 917–924.

Young, J.A., Fivelman, Q.L., Blair, P.L., De V, L., Le Roch, K.G., Zhou, Y., Carucci, D.J., Baker, D.A., and Winzeler, E.A. (2005), The *Plasmodium falciparum* sexual development transcriptome: A microarray analysis using ontology-based pattern identification, *Mol. Biochem. Parasitol.*, **143**, 67–79.

Chapter 2

Constructing Probabilistic Genetic Networks of *Plasmodium falciparum* from Dynamical Expression Signals of the Intraerythrocytic Development Cycle

Junior Barrera[a], Roberto M. Cesar Jr.[a], David C. Martins Jr.[a],
Ricardo Z.N. Vêncio[a], Emilio F. Merino[b], Márcio M. Yamamoto[b],
Florencia G. Leonardi[a], Carlos A. de B. Pereira[a] and
Hernando A. del Portillo[b]

[a]*Instituto de Matemáticas e Estatística e BIOINFO-USP,*
Núcleo de Pesquisas em Bioinformática, Universidade de São Paulo,
R. do Matão 1010, São Paulo, SP 05508-900, Brazil
[b]*Instituto De Ciências Biomédicas, Departamento de Parasitologia, Universidade de São*
Paulo, Ave. Lineu Prestes 1374, São Paulo, SP 05508-900, Brazil

Abstract   The completion of the genome sequence of *Plasmodium falciparum* revealed that close to 60% of the annotated genome corresponds to hypothetical proteins and that many genes, whose metabolic pathways or
biological products are known, have not been predicted from sequence
similarity searches. Recently, using global gene expression of the asexual
blood stages of *P. falciparum* at 1 h resolution scale and Discrete Fourier
Transform based techniques, it has been demonstrated that many genes
are regulated in a single periodic manner during the asexual blood stages.
Moreover, by ordering the genes according to the phase of expression,
a new list of targets for vaccine and drug development was generated.
In the present paper, genes are annotated under a different perspective:
a list of functional properties is attributed to networks of genes representing subsystems of the *P. falciparum* regulatory expression system.
The model developed to represent genetic networks, called Probabilistic
Genetic Network (PGN), is a Markov chain with some additional properties. This model mimics the properties of a gene as a non-linear stochastic gate and the systems are built by coupling of these gates. Moreover, a tool that integrates mining of dynamical expression signals by
PGN design techniques, different databases and biological knowledge,
was developed. The applicability of this tool for discovering gene networks of the malaria expression regulation system has been validated
using the glycolytic pathway as a "gold-standard", as well as by creating an apicoplast PGN network. Presently, we are tentatively improving
the network design technique before trying to validate results from the
apicoplast PGN network through reverse genetics approaches.

Keywords: malaria, annotation tool, probabilistic genetic networks, dynamical sys-
                                 tem, Markov chain, mutual information, gene expression, microarrays

1.   INTRODUCTION

Malaria remains the most devastating parasitic disease worldwide, and every year is responsible for 300–500 million clinical cases and 1–2 million deaths, mostly in children below 5 years old [http://www.who.int/tdr/diseases/malaria/default.htm]. Furthermore, the appearance of drug-resistant parasite strains to most antimalarial drugs and of insecticide-resistant *Anopheles* mosquitoes, in addition to the global warming, all have exacerbated this public health situation.

The advent of genomics into malarial research is significantly accelerating the discovery of control strategies. Indeed, the first draft of the complete genome sequence of *Plasmodium falciparum*, the most deadly human malaria parasite, was released only three years ago [9], but it has substantially modified the way of thinking for the development of new vaccines, drugs and alternatives of control strategies. Moreover, it has allowed the initiative for global scale studies on the transcriptome [1,4,10,12,16], proteome [7,10,14,15] and metabolome [23] of the parasite in different developmental stages.

Recent experimental evidence indicates that malaria parasites present unique mechanisms for control of gene expression: data from SAGE analysis has demonstrated that approximately 17% of abundant tags correspond to anti-sense transcripts of annotated genes [17], that suggests that these anti-sense transcripts might be involved in post-transcriptional regulation; reverse genetics approaches have shown that introns co-regulate expression of variant genes [2]; although promoters seem to be bi-partite, it is postulated that there must be unique sets of malarial transcription factors due to the high AT-content of intergenic regions and absence of recognized regulatory transcription factors [3,13].

Progressing the research effort, dynamical global gene expression measures of the intraerythrocytic developmental cycle (IDC) of the parasite at 1 h-scale resolution were recently reported [1]. Moreover, using Discrete Fourier Transform (DFT) based techniques, researchers verified that, during this life stage, the parasite seems to follow a rigid clockwise program where genes with common functions are transcribed at similar times. This study recognized 73% of the quality controlled (QC) dataset available for the CAMDA contest (http://www.camda.duke.edu/camda04/datasets/). The QC dataset comprises 3719 elements with relative expression signals with almost sinusoidal shape in the logarithmic scale or, equivalently, pulse like shape in the original relative expression scale. By ordering these signals by phase, they constructed a wave of signal propagation and ordered genes. Analysis of ordered genes throughout the asexual blood stages provided a comprehensive and biologically meaningful list of genes with putatively similar functions [1]. This analysis, however, did not include the elements that did not have almost sinusoidal shape and which, however, represented 27% of the QC dataset (i.e., 1361 elements).

In this paper, a list of functional properties is attributed to networks instead of individual genes. To do so, a tool was created that integrates mining procedures of dynamical expression signals and conventional databases (i.e., genome, proteome, metabolome, and clinical data).

This annotation approach may be applied to all spotted oligonucleotides of the QC set, despite the shape of their dynamical signals being sinusoidal or not. Subsystems of the malaria expression regulation system are modeled as probabilistic genetic networks (i.e., a stochastic process that is a specialization of a Markov chain) [20]. These networks are designed from the observed dynamical signals. The designed subsystems are annotated using conventional public databases and biological knowledge. The subsystems to be designed are defined from seed genes of particular biological interest, i.e. the subsystems are composed by genes that predict or are predicted by seed genes [11]. For example, some genes analyzed by the DFT approach were used as seeds to discover other non-sinusoidal genes associated to the same phase of the parasite life cycle.

Following this Introduction, Section 2 presents the concept of probabilistic genetic network (PGN). Section 3 describes the technique used for designing a PGN. Section 4 describes the developed software tools. Section 5 gives results of the application of the design techniques to simulated PGNs and presents preliminary biological results obtained by applying the proposed technique. Finally, the results and future steps of this research are discussed in the Concluding Remarks.

2. PROBABILISTIC GENETIC NETWORKS

The life of an organism depends on many metabolic pathways that are regulated by gene expression networks. The mechanism of pathways regulation involves a complex system with many forward and feedback signals. These signals are RNA, produced by gene expression, and protein complexes, produced by interaction of proteins built by translation of mRNA. Protein complexes act as feedback signals that control gene transcription. Forward signals, in the form of enzymes, act as control metabolic pathways. In such networks, the expression of each gene depends both on its own expression and on the expression levels of other genes at previous time instants. This complex network of interactions can thus be modeled by a dynamical system.

Finite dynamical systems, discrete in time and finite in range, can model the behavior of gene expression networks. In such model, we represent each transcript by a variable that takes the expression value of that transcript. All these variables, taken collectively, are the components of a vector called the *state of the system*. Each component (i.e. transcript) of the state vector has an associated function that calculates its next value (i.e. expression value) from the state at previous time instants. These functions are the components of a function vector, called *transition function*, which defines the transition from one state to the next and represents the gene regulation mechanisms. In order to formalize these ideas, we introduce some definitions and notation. Let R be the range of all state components. For example, $R = \{0, 1\}$ in binary systems, and $R = \{-1, 0, 1\}$ in three levels systems. The transition function ϕ, for a gene network of n genes, is a function from R^n to R^n. A finite dynamical system is given by

$$x[t + 1] = \phi(x[t]),$$

where $x[t] \in R^n$, for every $t \geqslant 0$. A component of $x[t]$ is a value $x_i[t] \in R$.

Systems defined as above are *time translation invariant*, i.e. the transition function is the same for all discrete time t. When ϕ is a stochastic function (i.e. for each state $x[t]$, the next state $\phi(x[t])$ is a realization of a random vector), the dynamical system is a stochastic process.

In this paper, we represent gene expression networks by stochastic processes. The stochastic transition function is a particular family of Markov chains called probabilistic genetic network (PGN).

Consider a sequence of random vectors $X_0, X_1, X_2, \ldots$ assuming values in R^n and its realizations denoted, respectively, $x[0], x[1], x[2], \ldots$. A sequence of random states $(X_t)_{t=0}^{\infty}$ is called a Markov chain if, for every $t \geqslant 1$,

$$P\big(X_t = x[t] | X_0 = x[0], \ldots, X_{t-1} = x[t - 1]\big)$$
$$= P\big(X_t = x[t] | X_{t-1} = x[t - 1]\big).$$

The significance of a Markov chain lies in the fact that the conditional probability of the future event, given the past history, depends only upon the immediate past and not upon the remote past.

A Markov chain is characterized by a transition matrix $\pi_{Y|X}$ of conditional probabilities between states, whose elements are denoted $p_{y|x}$, and an initial condition random vector of states π_0. The stochastic transition function ϕ at the time t is given by

$$\phi(x[t]) = y,$$

for every $t \geqslant 1$, where y is a realization of a random vector with distribution $p_{\cdot|x[t]}$.

A *Probabilistic Genetic Network* (PGN) is a Markov chain $(\pi_{Y|X}, \pi_0)$ such that

    (i) $\pi_{Y|X}$ is homogeneous, i.e. $p_{y|x}$ is not a function of t.
    (ii) $p_{y|x} > 0$, for every pair of states $x, y \in R^n$.
    (iii) $\pi_{Y|X}$ is conditionally independent, i.e. for every pair of states $x, y \in R^n$,

$$p_{y|x} = \prod_{i=1}^{n} p(y_i|x).$$

    (iv) $\pi_{Y|X}$ is almost deterministic, i.e. for every state $x \in R^n$, there exists a single state, $y \in R^n$ such that $p_{y|x} \approx 1$.
    (v) For every gene j there exists a vector a^j of integer numbers such that for every $x, z \in R^n$ and $y_j \in R$,

$$\text{if} \quad \sum_{i=1}^{n} a_i^j x_i = \sum_{i=1}^{n} a_i^j z_i \quad \text{then} \quad p(y_j|x) = p(y_j|z).$$

These axioms imply that each gene is characterized by a vector of coefficients a and a vector stochastic function g_j from Z, the set of integers numbers, to R. If a_i^j is positive then the target gene j is *excited* by gene i. If a_i^j is negative then it is *inhibited* by gene i. If a_i^j is 0, then it is *not affected* by gene i. We say that gene j is *predicted* by the gene i when a_i^j is different of 0. The component j of the stochastic transition function ϕ, denoted ϕ_j, is built by the composition of g_j with the linear combination of a^j and the previous state $x[t]$, i.e. for every $t \geqslant 1$,

$$\phi_j\big(x[t]\big) = g_j\left(\sum_{i=1}^{n} a_i^j x_i[t] \right),$$

where $g_j(\sum_{i=1}^{n} a_i^j x_i[t])$ is a realization of a random variable in R, with distribution $p(\cdot | \sum_{i=1}^{n} a_i^j x_i[t])$.

The axioms that define the PGN model are inspired in biological phenomena or mandatory simplifications due to the usual lack of data for the model estimation. The main hypothesis adopted is to choose a discrete model. This is justified because transcription and translation are discrete phenomena. The levels of quantization are chosen according to the available data for model parameters estimation.

Axiom (i) is a constraint just to simplify the estimation problem, but it could be generalized easily. Axiom (ii) imposes that all states are reachable, that is, noise may lead the system to any state. It is a quite general model that reflects our lack of knowledge about the kind of noise that may affect the system. Axiom (iii) means that the expression of a gene at a given time instant t does not depend on the expression of other genes at t. This happens when the time step of the model is less than the time spent for transcription-translation. Axiom (iv) means that the system has a main structural dynamics that is subject to small noise. This is what happens in practically all known engineering systems designed by man. Axiom (v) means that genes act as a non linear gate triggered by a balance between inhibitory and excitatory inputs, analogous to neurons.

It is important to recall that axiom (iii) might no be verified due to the 1h time resolution limitation of the available experimental data. However, this axiom was adopted in our model to allow statistical tractability. Of importance, using this axiom, we were able to generate biologically meaningful results (see below).

3.   DESIGN OF PGNS

The goal of this research is to estimate a PGN representing a subsystem of the malaria parasite gene expression network from dynamical microarray relative expression measures and biological knowledge. In the following the procedure adopted for PGN estimation is described.

The *entropy* $H(X)$ of a random variable X is a measure of its distribution $\{p_i\}$, given by

$$H(X) = -\sum_{i=1}^{n} p_i \log p_i.$$

The entropy has some remarkable properties: (i) all the distributions formed by permutations of p_i have the same entropy; (ii) concentrating the probability mass of a distribution implies in decreasing its entropy. As a corollary of property (ii), the uniform distribution presents maximum entropy and those with minimum entropy have the total probability mass concentrated in one point.

The *mutual information* [5] between two random variables X and Y is the measure defined by

$$I(X, Y) = H(Y) - H(Y|X).$$

It measures the probability mass concentration of $P(Y)$ in $P(Y|X)$ by the observation of X. The expectation $E[I(X, Y)]$ of $I(X, Y)$ is given by

$$E[I(X, Y)] = H(Y) - E[H(Y|X)].$$

When $E[I(X, Y)] = 0$, X and Y may be independent variables and the condition $P(Y) = P(Y|X)$ should be tested. In case this condition is true, then X and Y are independent, otherwise, they have dependence.

The expectation of the mutual information is used to estimate the PGN. The random variable Y will be the gene value $y_i[t+1]$ to be predicted and the given random variable X will be the vector of genes $x[t]$ weighted by an integer vector a, associated to gene y_i. For each vector a, with $a_i \in \{-1, 0, +1\}$ and at most three values different from 0, the mean mutual information is estimated. The first vectors a, that have larger mutual information, are selected. These vectors indicate the connection between genes and the kind of connection: excitatory or inhibitory.

4. DEVELOPED SOFTWARE TOOLS

The designed software system estimates gene networks from dynamical expression measures and represents them as graphs linked to malaria databases. Firstly, the system receives the raw fluorescence intensity measures as input and applies a quality control procedure that generates a new dataset. Then, the signals of this dataset are normalized and quantized into three expression levels $\{-1, 0, +1\}$.

Some target genes together with the quantized signals are provided to the main module of the system, which is responsible for computing the best predictors set for each target (based on the PGN design techniques described in the last section).

A user-friendly graphical interface was implemented to facilitate the biological interpretation of the results. The table of predictors, the file of functional groups annotated by Bozdech et al. [1] and the Overview dataset (http://www.camda.duke.edu/camda04/datasets) are organized and given as input for the GraphViz (a package to visualize graphs, http://www.research.att.com/sw/tools/graphviz). A color code was assigned to each node of the network (i.e. oligo) according to the functional biological categories defined in [1]: transcriptional machinery (pink), cytoplasmic translation machinery (blue), glycolitic pathways (yellow), etc. (see Figure 2). Besides, the node shape indicates if the oligonucleotide is present in the Overview set or not: a square indicates that it is present and a circle that it is not. Each node has a link to a page with pointers to three public databases: PlasmoDB (http://plasmodb.org),

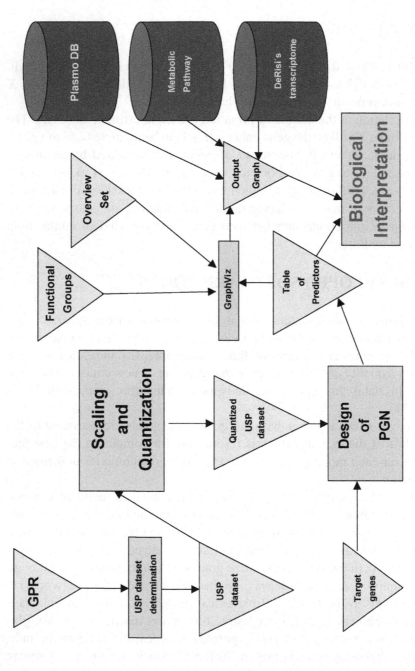

Figure 1. Simplified scheme of the analysis pipeline. Modules (squares) datafiles (triangles) databases (cylinders) and dataflow (arrows) of the data analysis pipeline.

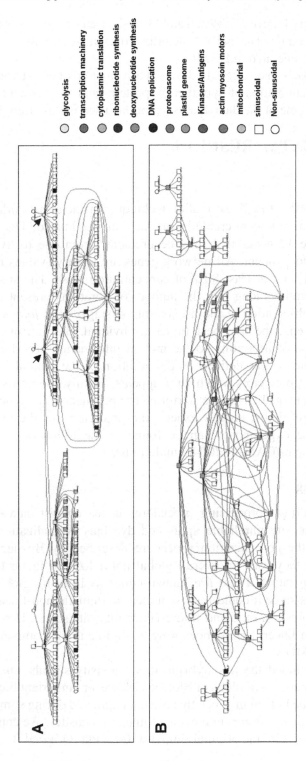

Figure 2. Glycolytic and Apicoplast Probabilistic Genetic Networks. Glycolytic (**A**) and Apicoplast (**B**) PGN networks were obtained by using those genes listed under functional group for glycolysis and plastid genome at the malaria IDC database (http://malaria.ucsf.edu/) as targets. Next, the five best tuples (individuals) of predictors were computed according to the mutual information criterion and a graph generated as described in Section 5.1. To facilitate visualization of the PGN networks, an arbitrary color-code was created to represent functional groups. Moreover, squares represent oligonucleotides with sinusoidal expression signals whereas circles represent oligonucleotides with non-sinusoidal expression signals. Glycolytic and Apicoplast networks can be visualized at http://www.vision.ime.usp.br/CAMDA2004/glycolysis.html and http://www.vision.ime.usp.br/CAMDA2004/apicoplast.html, respectively.

Metabolic Pathways (http://biocyc.org/PFA/) and DeRisi's transcriptome database (http://malaria.uscf.edu/). The output including the graph and links to public databases is fully generated in HTML.

Thus, this software allows easy access to different information of each target gene and can help in the annotation of hypothetical proteins and null elements. Figure 1 represents a scheme of the data analysis pipeline used in this study.

5.   EXPERIMENTAL RESULTS

5.1.   Simulations

For validating the proposed PGN estimation technique, artificial networks that satisfy the PGN definition were created, simulated and estimated. These simulated networks have 12 genes that may be predicted from one to five genes or may even be independent. All network genes are ternary (values in $\{-1, 0, +1\}$) and $p(y_i | X)$ has at least 80% of concentration mass. The simulations were just 48 iterations long (i.e. the number of iterations present at an 1 h-scale resolution observation of the asexual blood stages of *P. falciparum*). For each target gene, the five best tuples (individual, pairs, triples) of predictors were computed according to the mutual information criterion. The *quality of a predictor g* was defined as the addition of the mutual information of all tuples of predictors in which *g* appears. Finally, the predictors were ordered by their quality. In the performed experiments, the genes with greater quality were almost always exactly the predictors for the target gene. Some of these experiments can be found at the following site http://www.vision.ime.usp.br/CAMDA2004/simulations/.

5.2.   Pre-processing

We performed standard pre-processing procedures in the contest dataset such as filtering low-intensity unreliable spots and dye bias normalization. Moreover, we checked the normalization procedure described by Bozdech et al. [1] and found that they used an overall global normalization factor to normalize the expression ratios. There are known concerns in using global normalization procedures since it could represent clear systematic non-linear dependence between expression ratio and fluorescence intensities [18]. However, we verified that non-linear dependences were negligible in the complete dataset available for CAMDA.

Bozdech et al. [1] excluded the low hybridization intensity signals since they received the same treatment as blobs or blotches. However, important biological information may be hidden in genes that are not expressed during some part of the intraerythrocytic developmental cycle of malaria parasites. We constructed a different dataset from the original output of GenePix. Original flags for blobs or blotches were kept, whereas non-detectable expression values of low intensity signals were set to zero. We classified a spot as non-detectable if the mean intensity measurement in Cy3 or Cy5 is below to some local threshold value. This threshold is calculated from the distribution of pixel intensities of the background surrounding the spot. The 90% quantile of the local background distribution was used to define the intensity threshold. Spot's mean intensities below to this threshold were truncated to 0. This simple rule can naturally exclude unreliable signals, since the hybridization log-ratio $\log 2(0/0)$ is not defined. However, this rule preserves the potentially relevant situations when a signal is transcriptionally inactive only in a fraction of the time-course, since the expression becomes $\log 2(0/\text{reference}) = -\infty$. Although this is not a numerical ratio, the result can be incorporated in our Markovian approach because of the quantization step. As a result of this pre-processing step the USP-dataset used for the contest contained 6532 oligos, including 1361 oligos with not almost sinusoidal expression, as opposed to the 3719 oligos used in the overview dataset used by Bozdech et al. [1] to generate the phaseogram of the IDC malaria cycle.

5.3. Signal Normalization and Quantization

In order to validate the proposed methodology, the well known glycolytic pathway was studied. Before applying the predictor estimation techniques the signal was normalized and quantized. The signals were normalized by the *normal transformation* η given by, for every signal $g(t)$, $\eta[g(t)] = \frac{g(t) - E[g(t)]}{\sigma[g(t)]}$, where $E[g(t)]$ and $\sigma[g(t)]$ are, respectively, the expectation and standard deviation of $g(t)$.

The normal transformation has two important properties: (i) $E[\eta[g(t)]] = 0$ and $\sigma[\eta[g(t)]] = 1$, for every random variable $g(t)$; (ii) $\eta[g(t)] = \lambda \eta[g(t)]$, for every real number λ. The quantization of a gene at a given instant is a mapping from the continuous expression log-ratio into three qualitative expression levels $\{-1, 0, +1\}$ (i.e. down, null and up regulated in relation to the reference, respectively). The quantization of a gene signal g is performed by a *threshold mapping* given by

$$g'(t) = \begin{cases} +1 & \text{if } g(t) \geqslant h, \\ 0 & \text{if } l \leqslant g(t) \leqslant h, \\ -1 & \text{if } g(t) \leqslant l, \end{cases}$$

for every $t \geqslant 0$, where

$$l = \frac{\sum_{g(t)<0} g(t)}{|\{g(t): g(t) < 0\}|} \quad \text{and} \quad h = \frac{\sum_{g(t)>0} g(t)}{|\{g(t): g(t) > 0\}|}.$$

Normalization and quantization have the effect of creating equivalence classes between signals, thus decreasing estimation errors due to lack of data.

5.4.   Glycolytic PGN Network

During the asexual blood stages, malaria parasites rely entirely on glycolysis for its ATP production [21]. Thus, we chose target genes that code for all the 10 enzymes pertaining to the glycolytic pathway (hexokinase, phosphohexose isomerase, phosphofructokinase1, aldolase, triose phosphate isomerase, glyceraldehide 3 phosphate dehydrogenase, phosphoglycerate kinase, phosphoglycerated mutase, enolase, and pyruvate kinase) to test our model. Significantly, an interconnected glycolytic PGN network was generated by using the first five genes with the lowest entropic values associated to each glycolytic enzyme (Figure 2). Moreover, analysis of 40 best predictors for each glycolytic target (289 distinct oligos in total) revealed that most of them (96%) corresponded to hypothetical proteins genes related to transcription, translation, DNA and RNA synthesis, actin myosin motors and kinases (http://www.vision.ime.usp.br/CAMDA2004/Table1S.html). Remaining genes encoded surface antigens and thus, *a priori*, can be considered false-positives. Worth mentioning, similar results were obtained from a list of 400 genes expressed in-phase with glycolysis obtained from data of Bozdech et al. [1] (not shown). As expected, no genes of the TCA cycle were found in the glycolytic PGN network further corroborating the lack of a functional TCA cycle during the asexual blood stages of malaria parasites [21]. Of relevance, several oligonucleotides not included in the overview dataset due to low hybridization intensity or non-sinusoidal signals, were included in the PGN network (Figure 2). Of relevance, two oligonucleotides (opff72413 and m11919_1) corresponding to two glycolytic enzymes, hexokinase and aldolase, respectively, excluded from the phaseogram of Bozdech et al. [1], were included in the glycolitic PGN network. Together, this data demonstrates the value of the PGN model in generating a biologically meaningful glycolytic network that includes genes not included by the Fourier approach [1].

Next, we attempted to create an apicoplast PGN network. Enzymes from this organelle are becoming new targets for malaria since there is no homologous organelle in the human host [19,22]. Of relevance, two different computational algorithms have been developed to predict apicoplast proteins. In the first one, a genome-wide scan of *P. falciparum* revealed over 550 nuclear genes that encoded a consensus bi-partite peptide signal sequence [8]. In the second one, genes expressed in-phase with the plastid genome and containing the bi-partite peptide signal sequences narrowed the list of apicoplast nuclear-encoded genes from over 550 to 156 [1]. In order to apply our algorithm, oligonucleotides representing each of the 20 putative apicoplast genome-encoded proteins listed from DeRisi's laboratory (http://malaria.ucsf.edu/) were fed to our program and an apicoplast PGN network was generated (Figure 2). Analysis of the results clearly indicated that our method is capable of interconnecting genes that have been experimentally demonstrated to be part of the apicoplast (acyl-carrier protein and ribosomal protein S9), whereas many other genes lack predicted bi-partite peptide signal sequences. These results are difficult to reconcile with our present knowledge of the predicted malaria apicoplast proteome. Reverse genetics approaches similar to the ones used to define the importance of the bi-partite peptide signal sequences [8] can now be envisioned to validate some of these genes. Alternatively, our model describes genes not only related to the apicoplast proteome but genes whose expression is essential to create such network.

As our program creates PGN networks, a negative control was idealized to further validate the biological value of our findings. Thus, eight genes, four from glycolysis and four from the apicoplast organelle, were chosen randomly and used together as seed genes to create PGN networks based on single-gene and two-gene predictions. The results clearly demonstrated that the glycolysis and apicoplast PGN networks based on single-gene predictions were not interconnected (http://www.vision.ime.usp.br/CAMDA2004/ga.html). Based on two-gene predictions, with the exception of two genes from the glycolytic PGN network that inter-connected with the apicoplast PGN network, remaining genes were not connected (http://www.vision.ime.usp.br/CAMDA2004/ga2.html). It is important to recall that two-gene predictions are based on 21,330,246 calculations further reinforcing the value of these results. Together, this data demonstrates the value of the PGN model in generating biologically meaningful networks and which include genes not included by the Fourier approach [1].

An ideal PGN network will include interconnectivity networks based on interactions of several genes. Unfortunately, the "limited" amount of data presently available from the IDC transcriptome of *P. falciparum*, precludes such analyses without introducing a large degree of error. Regardless, this data demonstrates that the PGN model and program presented here are capable of constructing biologically meaningful networks of malaria from dynamical expression signals of the asexual blood stages and that it can be used as a complementary computational approach to Fourier analysis by including genes that are not periodically expressed.

6.   CONCLUDING REMARKS

In order to advance our knowledge on the biology of *P. falciparum*, we have designed PGNs from dynamical expression signals of the asexual blood stages reported by Bozdech et al. [1]. Unlike their DFT approach, PGN design allowed us to use all the elements available in the dataset. Significantly, this technique was applied to target genes that code for enzymes of the glycolytic pathway and a biologically meaningful glycolytic network was obtained. Next, we applied this algorithm to construct an apicoplast PGN network and although "signature" apicoplast genes were found, many other genes lack the consensus bipartite peptide signal sequence.

These results were obtained without considering the equivalence between linear combinations of inputs, what should improve the results, since the estimation errors will diminish and the hypothesis is quite consistent with observed gene dynamics. Besides, this model will permit to distinguish between inhibitory and excitatory signals. Although the normal transformation creates equivalence classes that diminishes the estimation errors, it amplifies noise in housekeeping genes that have almost constant expression signals. One way of circumventing this problem is to detect and exclude the housekeeping genes of the regulatory systems study before signal quantization.

The next steps of this research include mainly improving the network design technique and validation through reverse genetics approaches of some of the genes previously unpredicted by other algorithms as being part of the apicoplast. If validated, the PGN approach could thus be used to annotate genes not considered by the DFT approach and to accelerate the discovery of new targets against malaria.

ACKNOWLEDGEMENTS

Our thanks to Professor Bianca Zingales (IQ-USP) for first catalysing the interaction between JB and HAP, and to TDR-WHO for consolidating the interaction between the Departments of Parasitology and Computer Science through a Grant on Bioinformatics and Tropical Diseases (http://malariadb.ime.usp.br/courses.html). The authors also thank FAPESP (01/09401-0, 98/12765-8, 04/03967-0, 02/04698-8), CNPq and CAPES by their continuous support.

REFERENCES

[1] Bozdech, Z., Llinas, M., Pulliam, B.L., Wong, E.D., Zhu, J., and DeRisi, J.L., The transcriptome of the intraerythrocytic developmental cycle of *Plasmodium falciparum*, *PLoS Biol.*, 1 (2003), 5.

[2] Calderwood, M.S., Gannoun-Zaki, L., Wellems, T.E., and Deitsch, K.W., *Plasmodium falciparum* var genes are regulated by two regions with separate promoters, one upstream of the coding region and a second within the intron, *J. Biol. Chem.*, 278(36) (2003), 34125–34132.

[3] Coulson, R.M., Hall, N., and Ouzounis, C.A., Comparative genomics of transcriptional control in the human malaria parasite *Plasmodium falciparum, Genome Res.*, **14** (2004), 1548–1554.

[4] Daily, J.P., Le Roch, K.G., Sarr, O., Fang, X., Zhou, Y., Ndir, O., Mboup, S., Sultan, A., Winzeler, E.A., and Wirth, D.F., In vivo transcriptional profiling of *Plasmodium falciparum, Malar J.*, **3** (2004), 30.

[5] DeGroot, M.H., Uncertainty, information and sequential experiments, *Ann. Math. Statist.*, **3** (1962), 404–419.

[6] Dougherty, E.R., Bittner, M.L., Chen, Y., Kim, S., Sivakumar, K., Barrera, J., Meltzer, P., and Trent, J.M., In: *Proceedings of Nonlinear Filters in Genomic Control. IEEE-EURASI Workshop on Nonlinear Signal and Image Processing* (Antalia, Turkey, 1999), pp. 10–15.

[7] Florens, L., Washburn, M.P., Raine, J.D., Anthony, R.M., Grainger, M., Haynes, J.D., Moch, J.K., Muster, N., Sacci, J.B., Tabb, D.L., Witney, A.A., Wolters, D., Wu, Y., Gardner, M.J., Holder, A.A., Sinden, R.E., Yates, J.R., and Carucci, D.J., A proteomic view of the *Plasmodium falciparum* life cycle, *Nature*, **419** (2002), 520–526.

[8] Foth, B.J., Ralph, S.A., Tonkin, C.J., Struck, N.S., Fraunholz, M., Roos, D.S., Cowman, A.F., and McFadden, G.I., Dissecting apicoplast targeting in the malaria parasite *Plasmodium falciparum, Science*, **299** (2003), 705–708.

[9] Gardner, M.J., Hall, N., Fung, E., White, O., Berriman, M., Hyman, R.W., Carlton, J.M., Pain, A., Nelson, K.E., Bowman, S., Paulsen, I.T., James, K., Eisen, J.A., Rutherford, K., Salzberg, S.L., Craig, A., Kyes, S., Chan, M.S., Nene, V., Shallom, S.J., Suh, B., Peterson, J., Angiuoli, S., Pertea, M., Allen, J., Selengut, J., Haft, D., Mather, M.W., Vaidya, A.B., Martin, D.M., Fairlamb, A.H., Fraunholz, M.J., Roos, D.S., Ralph, S.A., McFadden, G.I., Cummings, L.M., Subramanian, G.M., Mungall, C., Venter, J.C., Carucci, D.J., Hoffman, S.L., Newbold, C., Davis, R.W., Fraser, C.M., and Barrell, B., Genome sequence of the human malaria parasite *Plasmodium falciparum, Nature*, **419** (2002), 498–511.

[10] Hall, N., Karras, M., Raine, J.D., Carlton, J.M., Kooij, T.W., Berriman, M., Florens, L., Janssen, C.S., Pain, A., Christophides, G.K., James, K., Rutherford, K., Harris, B., Harris, D., Churcher, C., Quail, M.A., Ormond, D., Doggett, J., Trueman, H.E., Mendoza, J., Bidwell, S.L., Rajandream, M.A., Carucci, D.J., Yates III, J.R., Kafatos, F.C., Janse, C.J., Barrell, B., Turner, C.M., Waters, A.P., and Sinden, R.E., A comprehensive survey of the *Plasmodium* life cycle by genomic, transcriptomic and proteomic analyses, *Science*, **307** (2005), 82–86.

[11] Hashimoto, R.F., Kim, S., Shmulevich, I., Zhang, W., Bittner, M.L., and Dougherty, E.R., Growing genetic regulatory networks from seed genes, *Bioinformatics*, **20** (2004), 1241–1247.

[12] Hayward, R.E., Derisi, J.L., Alfadhli, S., Kaslow, D.C., Brown, P.O., and Rathod, P.K., Shotgun DNA microarrays and stage-specific gene expression in *Plasmodium falciparum* malaria, *Mol. Microbiol.*, **35** (2000), 6–14.

[13] Horrocks, P., Dechering, K., and Lanzer, M., Control of gene expression in *Plasmodium falciparum, Mol. Biochem. Parasitol.*, **95** (1998), 171–181.

[14] Lasonder, E., Ishihama, Y., Andersen, J.S., Vermunt, A.M., Pain, A., Sauerwein, R.W., Eling, W.M., Hall, N., Waters, A.P., Stunnenberg, H.G., and Mann, M., Analysis of the *Plasmodium falciparum* proteome by high-accuracy mass spectrometry, *Nature*, **419** (2002), 537–542.

[15] Le Roch, K.G., Johnson, J.R., Florens, L., Zhou, Y., Santrosyan, A., Grainger, M., Yan, S.F., Williamson, K.C., Holder, A.A., Carucci, D.J., Yates, J.R., III, and Winzeler, E.A., Global analysis of transcript and protein levels across the *Plasmodium falciparum* life cycle, *Genome Res.*, **14** (2004), 2308–2318.

[16] Le Roch, K.G., Zhou, Y., Blair, P.L., Grainger, M., Moch, J.K., Haynes, J.D., De La Vega, P., Holder, A.A., Batalov, S., Carucci, D.J., and Winzeler, E.A., Discovery of gene function by expression profiling of the malaria parasite life cycle, *Science*, **301** (2003), 1503–1508.

[17] Patankar, S., Munasinghe, A., Shoaibi, A., Cummings, L.M., and Wirth, D.F., Serial analysis of gene expression in *Plasmodium falciparum* reveals the global expression profile of erythrocytic stages and the presence of anti-sense transcripts in the malarial parasite, *Mol. Biol. Cell.*, **12** (2001), 3114–3125.

[18] Quackenbush, J., Microarray data normalization and transformation, *Nat. Genetics*, **32** (2002), 496–501.

[19] Ralph, S.A., Van Dooren, G.G., Waller, R.F., Crawford, M.J., Fraunholz, M.J., Foth, B.J., Tonkin, C.J., Roos, D.S., and McFadden, G.I., Metabolic maps and functions of the *Plasmodium falciparum* apicoplast, *Nat. Rev. Microbiol.*, **2** (2004), 203–216.

[20] Shimulevich, I., Dougherty, E.R., Kim, S., and Zhang, W., Probabilistic Boolean networks: A rule-based uncertainty model for gene regulatory networks, *Bioinformatics*, **18**(2) (2002), 261–274.

[21] Sherman, I.W., Metabolism and surface transport of parasitized erythrocytes in malaria, *Ciba Found Symp.*, **94** (1983), 206–221.

[22] Wilson, R.J.M. (Iain), Progress with parasite plastids, *J. Mol. Biol.*, **319** (2002), 257–274.

[23] Yeh, I., Hanekamp, T., Tsoka, S., Karp, P.D., and Altman, R.B., Computational analysis of *Plasmodium falciparum* metabolism: Organizing genomic information to facilitate drug discovery, *Genome Res.*, **14** (2004), 917–924.

Chapter 3

Simple Methods for Peak and Valley Detection in Time Series Microarray Data

A. Sboner[a,b], A. Romanel[b], A. Malossini[b], F. Ciocchetta[a,b], F. Demichelis[a,b], I. Azzini[a], E. Blanzieri[b] and R. Dell'Anna[a]

[a]*Bioinformatics Group, SRA Division, ITC-irst, Via Sommarive 18, I-38050 Povo (TN), Italy*
[b]*Department of Information and Communication Technology, University of Trento, Via Sommarive 14, I-38050 Povo (TN), Italy*

Abstract

Given a set of gene expression time series obtained by a microarray experiment, this work proposes a novel quality control procedure that exploits six analytical methods, each of which allows for the identification in an automated way of genes that have expression spikes within narrow time-windows and over a chosen amplitude threshold. The output of these methods, suitably combined in an automated way, provides an exhaustive list of genes and time points in which abrupt variations have been detected. The quality control on these genes is then performed by a biologist, who classifies the spikes either as biologically relevant or as artifacts. In the latter case, spikes must be eliminated by a smoothing procedure. In this chapter, we first describe the six methods and their iterative and automated implementation. As a case study, we discuss the application of the panel of these six methods to the transcriptome of *Plasmodium falciparum* intraerythrocytic developmental cycle. Assuming that spikes detected in this set have been labeled as artifacts by a biologist, in the second part of the chapter we discuss the effect of our smoothing procedure for different types of data analysis.

Keywords: malaria, DNA microarray, discrete mathematics, support vector machine (SVM), quality control

1.   INTRODUCTION

To develop new drugs and vaccines that disable the malaria parasite *Plasmodium falciparum* (*P. falciparum*) [19], researchers need a better understanding of the regulatory mechanisms that drive the malarial life cycle. In [2], the first comprehensive transcriptome analysis of the *P. falciparum* asexual cycle, or intraerythrocytic developmental cycle (IDC), which is associated with the clinical symptoms of malaria, is provided. Data in [2] show that: (1) at least

60% of the genome is transcriptionally active during this stage, and (2) *P. fal-ciparum* has developed an extremely specialized mode of transcriptional regulation. A continuous cascade of gene expression is produced, beginning with genes corresponding to general cellular processes, and ending with *Plasmod-ium*-specific functionalities, most of which are poorly understood. In other recent works on the biology of *P. falciparum* [3,15], attention is mainly devoted to the poor knowledge of the *P. falciparum* gene functionalities. In fact, the malaria genome sequencing consortium estimates that more than 60% of the 5,409 predicted open reading frames (ORFs) lack sequence similarity with genes from any other known organism [8].

The simple program regulating the life of *P. falciparum* may hold the key to its downfall, as any perturbation of the regulatory program may have harmful consequences for the parasite [20]. The simple cascade of gene regulation that directs the asexual development of *P. falciparum* is unprecedented in eukaryotic biology [2]. The transcriptome of the IDC resembles a "just-in-time" manufacturing process, whereby induction of any given gene occurs once per cycle and only at specific time points when required [2].

Quality control in microarray data analysis aims at discarding flawed data at an early stage of the analysis. The typical quality control procedure is performed after measurements on the raw digital image, in order to ensure that the measurements are not affected by image artifacts and thus increasing signal-to-noise ratio. However, given the experimental structure of present datasets, namely the time series component, it is possible to use such temporal information in order to further detect expression points that could be still affected by noise. Abrupt variations in the transcriptional profile can indicate anomalous behavior that needs to be assessed by a biologist as (a) being artifacts, or (b) carrying relevant biological information. Among abrupt variations, we were particularly interested in peaks and valleys, as they preserve signal periodicity, which (as shown in [2]) is an IDC transcriptome characteristic. Usually, time-series analysis [1,5,6] first approximates temporal signals by a continuous interpolating function. However, in this study we chose to preserve the actual information contained in each time point. In fact, our goal is to identify ORFs that show a relevant variation with very short duration with respect to the overall length of IDC (48 hours). To achieve this goal, we set up five different simple methods based on the discrete derivative and integral operators. An additional sixth method directly matches abrupt variations on transcriptional profiles. These six methods separately perform gene expression investigation in an iterative and automated way, thus avoiding time-expensive, direct visual inspection of all available time series microarray data. The output lists of detected genes and time points from the six methods generally overlap but are not coincident because the six methods are concerned with different behaviors in the temporal signal. Therefore, by merging in an automated way the six different output lists (see Section 3.1), a more comprehensive list of genes and time points at which relevant peaks and valleys are present can be obtained. As mentioned, the detected spikes can be classified by the biologist either as biologically relevant or as artifacts. In the former case, further analysis for biological interpretation of the results is required. In the latter case, peaks and valleys are artifacts that were not detected by conventional quality control procedures. They are therefore removed and substituted by a smoothing procedure that preserves the periodic nature of the overall signal.

The first part of the chapter is devoted to the description of our procedure for peak and valley detection. We discuss the application of the panel of the six methods to the transcriptome of *P. falciparum* IDC. Assuming that any biological relevance of the abrupt variations in this set is ruled out by a biologist, in the second part of the chapter we check whether the smoothing procedure influences further analysis. We find that our smoothing procedure changes the data analysis results. Our quality control procedure can therefore be effective either in further improving the signal to noise ratio of time series microarrays data or in highlighting possible biologically important time points in the same data.

2.   PRELIMINARY ANALYSIS

Given the intrinsic complexity of the experiments involving DNA microarray (see for example [10,17]), we investigated thoroughly the reliability of the contest datasets [2,4]. In particular, on a selected sample set of available data, we: (1) performed a visual inspection of microarray images (the "Primary Data" in [4]), (2) used TIGR SpotFinder [12] to analyze these images, and finally, (3) checked the results of GenePixPro3.0 quality control algorithm. GenePixPro3.0 [9] is the software used in [2] to acquire and analyze the DNA microarray data. The results and considerations obtained from this step of our work suggested us to use the "QC_dataset" [2,4]. This is the set of oligonucleotides that passed all quality control filters and was obtained from the "Complete_Dataset" [2,4]. This choice presents some positive aspects: oligos with many missing data, which may affect the results of our methods, are not present; gene expression values obtained from corrupted images are also not included. Moreover, this choice allows us to prove that our quality control procedure is able to further increase the signal to noise ratio. The "QC_dataset" contains 5080 of the 7091 oligonucleotides provided by Bozdech et al. [2].

3.   METHODS OF ANALYSIS

3.1.   Detection Methods

Following [13], we considered the "QC_dataset" as the matrix depicted in Table 1. We label this matrix **E**, denoting with $E(o, t)$ an element of **E**. The variable o indexes the oligos from $Oligo_1$ to $Oligo_{5080}$, and for the variable t, $t \in TP$, where $TP = \{TP_1, \ldots, TP_{22}, TP_{24}, \ldots, TP_{28}, TP_{30}, \ldots, TP_{48}\}$. TP_{23} and TP_{29} were not provided by [2,4]. Missing values in Table 1 were imputed with the "loess()" local regression function, provided by the "stats" package of R (version 2.01) [11,2]. The local weighting parameter was reduced to 12%.

In order to find within the **E** matrix gene expressions with rapid changes in time (in particular, candidate peaks and valleys), we exploited six different methods (labeled M_i, $i = 1, \ldots, 6$), concisely reported in Table 2. They can be split into three main classes: derivative methods (M_1, M_2, M_3), integral methods (M_4, M_5), and other methods (M_6).

3.1.1.   Method Description.   Each method can be described at abstract level as follows. For each transcriptional profile, method M_i detects a time point τ_i in which the expression variation occurs. This is accomplished by means of the score S_o. For methods M_1, M_2, M_4, M_5 and M_6, the higher the score S_o, the higher the probability to find a significant peak (or valley) with respect to the average signal amplitude. Contrary to the other five methods, for method M_3 the closer to zero is S_o, the higher is the probability to find a significant peak (or valley).

All methods M_i differ in the way they calculate S_o. Method M_1 proceeds for each oligo o as described in Figure 1:

Step 1: M_1 computes at each time point t the discrete derivative, calculated as the ratio of finite differences of width one;

Step 2: the maximum absolute value of the discrete derivative of Step 1 is calculated. This value is S_o.

The same procedure characterizes method M_2 with the discrete derivative calculated as the ratio of finite differences of width two.

Method M_3 proceeds for each oligo o as follows:

Table 1. The data matrix **E** obtained by QC_dataset

Oligo	TP_1	...	TP_{48}
$Oligo_1$	$\log_2(Cy5/Cy3)$	...	$\log_2(Cy5/Cy3)$
...	...	...	...
$Oligo_{5080}$	$\log_2(Cy5/Cy3)$	...	$\log_2(Cy5/Cy3)$

Given in input matrix **E**,
  Do ∀ oligo o,

  {

   Step 1. (Discrete derivative). $\forall t \in$ TP compute:
   $$\frac{\Delta E(o, t)}{\Delta t} = \frac{E(o, t+1) - E(o, t)}{(t+1) - t} = \Delta E(o, t)$$
   Step 2. (Score). Compute:
   $$\max_{t \in TP} |\Delta E(o, t)| = S_o$$

  }

Figure 1. The derivative method M_1.

Step 1: the discrete derivative $\Delta E(o, t)$ is calculated at each time point t, as in M_1;

Step 2: the maximum and minimum values of $\Delta E(o, t)$ are calculated;

Step 3: S_o is calculated as reported in Table 2. As already pointed out, the smaller is S_o, the higher is the probability to find a significant peak (or valley). This formula allows us to discriminate between spikes and unit-step like behavior of the signal.

Method M_4 is reported in Figure 2. For each oligo o it proceeds as follows:

Step 1: a normalization is performed, by subtracting from each element $E(o, t)$ the arithmetic mean computed on the whole temporal signal;

Step 2: the discrete integral A_1 of the absolute value of the normalized signal is calculated;

Step 3: the discrete derivative (width one) $\Delta E(o, t)$ of the original signal is calculated;

Step 4: the maximum value of $\Delta E(o, t)$ is calculated and the time point τ at which it occurs is stored;

Step 5: the positive integral A_2 of the normalized signal of width two around τ is calculated;

Step 6: the fraction of area $S_o = A_2/A_1$ is calculated.

The same procedure as in M_4 characterizes method M_5, except for Step 4, in which the minimum value of $\Delta E(o, t)$ is calculated. In other words, M_4 detects peaks while M_5 detects valleys.

Methods M_6, reported in Figure 3, proceeds for each oligo o as follows:

Step 1: time points t, $t+1$, $t+2$ are considered (for each $t \in$ TP) and the values α and β are calculated (see Figure 3). In case of a perfect spike, $\alpha = \beta$. For a first type discontinuity (a unit-step function) α or β is zero. There-

Given in input Matrix **E**,

    Do $\forall$ oligo o,

      {

        Step 1. (Normalization). Compute: $\forall t \in TP$

$$\overline{E}(o,t) = E(o,t) - \underset{t \in TP}{\mathrm{mean}}(E(o,t))$$

        where *mean* is the arithmetic mean over time

        Step 2. (Integral). Compute:

$$\sum_{t \in TP} \left| \overline{E}(o,t) \right| = A_1$$

        Step 3. (Discrete derivate), compute:

$$\frac{\Delta E(o,t)}{\Delta t} = \frac{E(o,t+1) - E(o,t)}{(t+1) - t} = \Delta E(o,t)$$

        Step 4. (Maximum localization). Find:

$$\tau = \arg \max_{t \in TP} \left(\Delta E(o,t) \right)$$

        Step 5. (Local integral). Compute:

$$\sum_{t = \tau - 1}^{\tau + 1} \left| \bar{E}(o,t) \right| = A_2$$

        Step 6. (Score). Compute $\dfrac{A_2}{A_1} = S_o$

      }

Figure 2. The integral method M_4.

fore $SV(E(o,t))$ yields one in the first case, and zero in the second. The term $(\alpha + \beta)/2$ weights the asymmetry of non perfect spikes;

    Step 2: the properly normalized maximum value of $SV(E(o,t))$ gives score S_o.

Method M_6, therefore, looks for three-point structures in each gene profile, weighting their possible asymmetry and selecting that structure for which the area is maximal.

3.1.2.   **Spike Detection.** In order to single out peaks and valleys in temporal signal, an amplitude threshold, called *pv*, must be given. The spike detection procedure identifies in an automated way those expression variations which are greater than *pv*. In this procedure, each method M_i is separately and

> Given in input Matrix **E**,
>
> Do ∀ oligo *o*,
>
> {
>
>   **Step 1.** (Spike value detection). $\forall t \in \text{TP}$ compute:
>
> $$\alpha = \left| E(o, t+1) - E(o, t) \right|$$
> $$\beta = \left| E(o, t+2) - E(o, t+1) \right|$$
> $$SV(E(o, t)) = \frac{\min\{\alpha, \beta\}}{\max\{\alpha, \beta\}} \cdot \frac{\alpha + \beta}{2}$$
>
>   **Step 2.** (Score). Compute:
>
> $$\frac{\max_{t \in \text{TP}} SV(E(o, t))}{\sum_{t \in \text{TP}} SV(E(o, t))} = S_o$$
>
> }

Figure 3. The method M_6.

Table 2. The methods M_i for spike detection

Methods	Description
Derivative	
M_1	Figure 1
M_2	As M_1, with Step 1 in Figure 1 replaced by:
	$\frac{\Delta E(o,t)}{\Delta t} = \frac{E(o,t+2)-E(o,t)}{(t+2)-t} = \Delta_2 E(o,t)$
M_3	As M_1, with Step 2 in Figure 1 replaced by:
	$[\,\mid \max\limits_{t \in \text{TP}} (\Delta E(o,t)) \mid - \mid \min\limits_{t \in \text{TP}} (\Delta E(o,t)) \mid] - [\max\limits_{t \in \text{TP}} (\Delta E(o,t)) - \min\limits_{t \in \text{TP}} (\Delta E(o,t))] = S_o$
Integral	
M_4	Figure 2
M_5	As M_4, with "argmax" replaced by "argmin" (Step 3, Figure 2)
Other methods	
M_6	Figure 3

iteratively applied to each expressionary time series of the considered dataset. Each method M_i provides its final list of genes and time points at which the expression values are greater than pv. In other words, for each method M_i the iterative procedure can be schematized as follows: (i) For each expressionary time series $E(o, t)$, the time point τ_i is found for which the maximum value of S_o occurs (or minimum value S_o for method M_3); (ii) if the expression value $E(o, \tau_i)$ is greater than pv, $E(o, \tau_i)$ is substituted by applying in τ_i the "loess()" function with local weighting parameter reduced to 15%, and the value τ_i is stored; (iii) steps (i) and (ii) are repeated until no new τ_i is found in which the expression value is greater than pv; (iv) a set of oligos for which at least one spike is found, and the list of the corresponding time points τ_i ($i = 1, \ldots, k$ with k = number of detected spikes) is provided as output. This iterative and automated procedure for each method M_i is implemented in R [11].

The lists of oligos and time points provided by the six methods are not necessarily the same, as S_o is differently calculated for each method. Therefore, a more comprehensive list is obtained by merging the six contributes and discarding redundancies in the merged list. This combination is performed in an automatic way.

The final number of affected oligos and of detected time points depends on the threshold value (pv) chosen. Therefore, the spike detection is carried out for different pv values. A small value of pv (with respect to the average signal amplitude) does not allow us to discriminate between simple amplitude fluctuations and abrupt variations, while too large a pv value may miss spikes which could be relevant for the quality control procedure (see as an example Section 4). Therefore, by analyzing the obtained numbers and performing a visual inspection of the correspondingly identified expression profiles, the most appropriate pv value can be chosen (see as an example Table 3, Figures 4.A1 and 4.B1).

3.1.3.   Smoothing Procedure.

At this point, the list of oligos and time points related to the chosen pv value is passed on to a biologist. If he/she does not assign any biological importance to the detected peaks and valleys, the expressionary time series of the original dataset carrying those spikes are substituted by the corresponding smoothed profiles. These smoothed profiles are obtained by applying the "loess()" function, with local weighting parameter reduced to 30%, to each previously detected time point. Figure 4 presents two examples of expression time series identified by the iterative procedure, performed with pv equal to 2. Expression data before and after the described smoothing are reported therein.

3.2.   Evaluation of the Detection Methods

In the case of artifact detection, it is necessary to provide evaluation methods in order to assess the impact that their smoothing-out can have on further analysis. In other words, it is necessary to check if the smoothing procedure has some effect on the results of data analysis. We considered a functional classification with support vector machine and the power spectrum analysis.

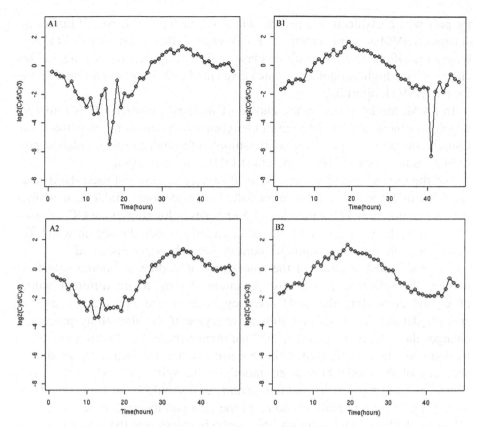

Figure 4. Example of expression time series detected by the iterative procedure, performed with $pv = 2$. Profiles before (A1 and B1 panels) and after (A2 and B2) the described smoothing are reported. Genes reported in A and B are respectively b541 (detected time points: TP_{15}, TP_{16}, TP_{17}, TP_{18}) and f71224_1 (detected time points: TP_{39}, TP_{40}, TP_{41}, TP_{43}).

In the following, we discuss the application of our quality control procedure to the QC_dataset described in [2,4]. The application of the spike detection method to the QC_dataset, and the subsequent evaluation performed by biology experts, led to the smoothing of the original expressionary time series in the time points detected by the methods. Therefore a new dataset, QC_dataset_smooth, was built.

3.2.1. Effect of Spike Smoothing on a MSVM Functional Classification.
Support vector machine (SVM) is a state-of-the-art classifier which has been widely used in the analysis of microarray data [7,14,18]. We studied the effect of spike smoothing on a multi-class SVM (MSVM) classifier [13] provided by the package "e1071" of R [11] by considering its influence both on model selection and on functional class prediction. In particular, we adopted

the pair-wise classification approach, where for each possible pair of functional classes an SVM classifier is trained. For N classes, this results in $(N - 1) \cdot N/2$ binary classifiers, and the resulting class is chosen by majority voting, i.e. the class with the highest number of votes gives the label. We chose a linear kernel for the MSVM algorithm.

In SVM, model selection the choice of the cost parameter C is required, which sets the trade-off between model complexity and generalization error. Usually, the best cost parameter C is estimated through a cross validation procedure, as in our case a leave-one-out (LOO) cross validation.

For the model selection analysis, as a training set we first used the dataset provided in TableS2 [2,4], hereafter called "raw_dataset". TableS2 describes the known functional classification of 530 genes belonging to the QC_dataset. Afterwards, the second dataset, hereafter called smooth_dataset, in which the same genes are extracted from QC_dataset_smooth, was considered.

We evaluated the effect of the smoothing procedure on model selection, namely the selection of the cost parameter C, by fixing different values of C and computing the LOO accuracy both on the raw_dataset and the smooth_dataset. This analysis aims at verifying if the smoothing procedure changes the LOO accuracy on each C parameter, therefore affecting model selection and thus classification task. Our aim is not at evaluating the predictive accuracy of the model built after smoothing the spikes, for which the small degree of bias (many cases and few parameters) should be instead considered as in [21]. Assuming that the choice of the cost parameter C is performed by selecting the best LOO accuracy, this analysis shows that the smoothing procedure will lead the experimenter to choose a different value of C, and thus different MSVM models for the subsequent analysis.

Afterwards, to assess the impact of the smoothing procedure on functional classification, we trained the MSVM on the raw_dataset and on the smooth_dataset, and used the obtained models to predict the genes without functional annotation in the QC_dataset, and in the QC_dataset_smooth, respectively. Given the raw_dataset and the smooth_dataset, in both cases the parameter C maximizing LOO accuracy was chosen to build the corresponding model. Any two C parameters showing similar LOO accuracy could be used instead. In fact, we do not aim at selecting the two models with highest predictive accuracy; but we want to point out the differences in the functional classification of the two MSVM models due to our smoothing procedure. The idea is to isolate the effects of the smoothing procedure on the functional classification results. However, our choice for the C parameters is that corresponding to the maximum LOO accuracy, thus resembling the standard choice for model selection.

The result shows that even two MSVM models showing similar LOO accuracy classify differently the unknown oligos. We do not know which is the "right" classification, but we point out that the smoothing procedure plays a great role for the subsequent biological investigations.

3.2.2.   Effects on Power Spectrum. We assessed how our smoothing procedure affects the power spectrum used in [2] to select the genes that have a definitely periodic time course. We thus repeated the computational steps therein described to obtain the power spectrum, using the QC_dataset as well as the QC_dataset_smooth, and compared the differences.

4.   RESULTS

Table 3 reports, for different values of *pv*, the number of oligos having at least one spike. For each value of *pv*, the list of oligos and time points were obtained as described in Section 3.1 by merging the results obtained by the different six methods M_i and considering only once the time points which are identified highlighted by more than one method.

Table 3 illustrates that the value *pv* equal to 2 sensibly discriminates between irrelevant time variations (*pv* < 2) and too-stringent spike-detection conditions (*pv* > 2). This choice was confirmed by visually inspecting a number of selected expressionary profiles, as those reported in Figures 4.A1 and 4.B1. As reported in Table 3, for *pv* = 2 the automated procedure identified 334 oligos, each presenting abrupt expression variation in at least one time point. Accordingly, a new dataset, "QC_dataset_smooth", was obtained by substituting in the original QC_dataset the 334 transcriptional profiles obtained by our procedure with their smoothed version. For the sake of simplicity, Table 4 only reports those 56 genes with the functional annotation. The complete list is available upon request.

Table 3. Number of oligos with at least
one spike detected by the iterative
procedure for different values of *pv*

pv	# oligos
1	3305
2	334
3	28
4	8
5	2
6	1
7	0

Table 4. Genes with functional annotation which present at least one detected spike in their expression (see Supplemental table for class acronym definition)

oligo_ID	Class	oligo_ID	Class	Oligo_ID	Class	oligo_ID	Class
a10325_30	ER	f739_1	MI	l1_28	ER	opfblob0060	AM
a10325_32	ER	i10472_1	MI	m14235_3	CT	opfblob0092	MI
a12696_3	MI	i1225_2	MI	m33088_2	AM	opfk12894	ER
a1718_1	DR	i14975_1	MI	m36656_1	MI	opfl0013	AM
b218	MI	i8675_1	AM	m54626_4	CT	opfl0022	AM
b230	MI	j116_7	MI	m60464_2	MI	opfl0029	M
b391	OT	j170_10	MI	n131_10	OT	opfl0141	AM
b444	MI	kn9335_1	DR	n132_124	MI	opfm60467	MI
d49942_9	MI	kn973_2	DR	n132_125	MI	ptrgln	PG
e15509_11	AM	ks1030_4	OT	n134_78	DR	ptrgly	PG
e18550_1	MI	ks26_17	AM	n137_2	CT	z_4_50	MI
e24991_1	MI	ks48_18	ER	n138_34	M	z_4_50	MI
f12313_1	MI	ks510_10	MI	n141_14	MI		
f27464_2	OT	ks510_8	MI	opfb0671	MI		
f49857_1	MI	ks75_15	ER	opfblob0020	ER		

We first assessed the distribution of those time points by computing the histogram reported in Figure 5. We can note that the methods identified more than 100 spikes at time point 18.

In this section we first discuss how the smoothing procedure affects MSVM model selection and functional classification. Consistency considerations are also reported.

Concerning the model selection, Table 5 reports the LOO accuracy values for different values of the cost parameter C.

From Table 5 it is evident that selecting the parameter with maximal LOO accuracy, using the raw_dataset the best parameter should be $C = 0.1$, while using the smooth_dataset the chosen parameter should be $C = 1.0$. Hence, despite of the very few modifications induced by the smoothing procedure (56 out of 530 genes of the training set), two different models should be selected.

We then predicted the functional class of genes without annotation in the QC_dataset as well as in the QC_dataset_smooth, as described in Section 3.2.1. In the confusion matrix we obtained 970 off-diagonal elements (out of 4550), i.e. 970 elements were classified differently by the two classifier obtained by raw_dataset and smooth_dataset. The confusion matrix regarding the prediction of functional expression of unknown genes between the two MSVM models, selected for $C = 0.1$ and $C = 1.0$ respectively, is provided as supplemental material.

Concerning the power spectrum analysis, the smoothing procedure, by eliminating abrupt changes in the signal, removes high frequency components in the Fourier space. Therefore, as expected, the power spectrum shifts towards

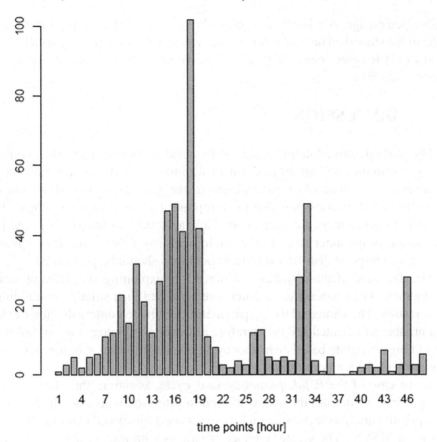

Figure 5. Spike temporal distribution.

Table 5. LOO accuracy values of MSVM for different values of cost parameter *C* using raw_dataset and smooth_dataset

C	LOO accuracy raw	LOO accuracy smoothed
0.001	56.4	56.8
0.01	69.4	69.6
0.1	72.5	71.9
1	72.1	72.5
10	69.2	69.4
100	67.4	67.4
1000	67.0	64.3
10000	64.7	66.8
100000	66.6	66.8

higher percentage. About 50 more genes have a power spectrum greater than 90% in the smoothed dataset. Concerning the cut-off value of 70%, which was used in [2] to select periodic genes, 12 more genes have a power spectrum greater than 70%.

5.   DISCUSSION

The study described in this paper can be divided into two conceptually distinct parts. In the first part we perform an automated quality control procedure by detecting anomalously rapid changes in the gene expression time series. Biologists need to assess whether these represent artifacts or are biologically relevant. In the former case, such anomalous rapid changes have to be properly accounted. In the latter case, further biological investigation on those spikes and on the temporal distribution of their positions should be performed.

The detection of these spikes is achieved by exploiting six different simple methods in an automated and iterative way, and then suitably combining their results. The choice of the *pv* parameter permits to control the amplitude and number of detected spikes, therefore allowing the biologist to control the smoothing procedure based on his/her own personal knowledge of the expected dynamics of the temporal series.

In the case of the *P. falciparum* asexual cycle, assuming that these peaks are artifacts, we discuss the effects of their substitution with smoothed values on a popular analysis technique such as supervised functional classification by means of MSVM. The greater number of valleys with respect to peaks seems to indicate that they are artifacts. In fact, in the case of low signals the relative noise is higher, so it seems reasonable to detect more valleys than peaks. We found that removing artifacts detected by our methods affects both the results of the MSVM model selection procedure and the MSVM functional classification of genes without annotation. In the latter case, 970 genes are differently classified before and after the smoothing procedure. It is worth noting that we do not discuss the degree of reliability of either classification. Our aim is to show that our quality control procedure influences data analysis results. It is also worth noting that the smoothing procedure we propose is locally applied only to the temporal points in which artifacts occur. Therefore, it preserves the overall temporal profile. This strengthens the effectiveness of our quality control procedure.

Concerning power spectrum computation, the smoothing procedure confirms and enhances the periodicity of the expression profiles used for subsequent analysis in [2]. This result is consistent with the aim of our quality control procedure at preserving as much as possible signal periodicity. However, though preserving periodicity, our approach may affect functional analysis.

In the temporal distribution of spike positions as reported in Figure 5, the most crowded channel is located at time steps 18. The result of Kolmogorov–Smirnov test performed on this distribution allows us to state with a high level of confidence ($p < 0.0005$) that this spike position distribution does not come from a uniform distribution, suggesting that spikes, if considered artifacts, are not due to random experimental errors. This analysis may suggest to biologists, aware of the performed experimental procedure, the possible causes of artifacts. In this way, improvements of the experimental process could be achieved.

Supplemental table. Confusion matrix regarding the prediction of functional expression of unknown genes between the two MSVM models, selected for $C = 0.1$ and $C = 1.0$, respectively

		MSVM with smooth_dataset $C = 1.0$													
		AM	CT	DR	DS	ER	GP	M	MI	OT	P	PG	RS	TC	TM
	AM	4	0	0	0	0	0	0	0	0	0	0	0	0	0
	CT	0	1393	0	0	7	31	0	0	0	33	0	75	0	27
	DR	0	0	340	7	0	0	50	1	103	0	3	0	21	0
	DS	0	0	0	0	0	0	0	0	0	0	0	0	0	0
	ER	1	27	0	0	181	0	0	0	0	0	0	2	0	4
	GP	0	14	0	0	0	51	0	0	1	29	0	20	0	1
MSVM	M	0	0	9	1	0	0	25	6	0	1	0	0	21	0
with	MI	62	46	7	0	5	1	3	654	4	16	2	1	16	0
raw_dataset	OT	0	10	54	1	0	1	10	0	349	8	0	15	15	0
$C = 0.1$	P	0	36	9	0	0	5	5	9	77	443	2	3	7	0
	PG	0	0	16	1	0	0	5	1	0	0	8	0	1	0
	RS	0	2	0	0	0	5	0	0	6	4	0	129	0	0
	TC	0	0	0	0	0	0	0	0	0	0	0	0	1	0
	TM	0	2	0	0	0	0	0	0	0	0	0	2	0	0

AM = Actin myosin motors, CT = Cytoplasmic Translation machinery, DR = DNA replication, DS = Deoxynucleotide synthesis, ER = Early ring transcripts, GP = Glycolytic pathway, M = Mitochondrial, MI = Merozoite Invasion, OT = Organellar Translation machinery, P = Proteasome, PG = Plastid genome, RS = Ribonucleotide synthesis, TC = TCA cycle, TM = Transcription machinery.

REFERENCES

[1] Bar-Joseph, Z., Analyzing time series gene expression data, *Bioinformatics*, **20**(16) (2004), 2493–2503.

[2] Bozdech, Z., Llinas, M., Pulliam, B.L., Wong, E.D., Zhu, J., and DeRisi, J.L., The transcriptome of the intraerythrocytic developmental cycle of *Plasmodium falciparum*, *PLoS Biol.*, **1**(1) (2003 October), e5 DOI: 10.1371/journal.pbio.0000005.

[3] Broudy, T., The modern age of malaria research: Finding new ways to combat an old disease, *Affimetrix Research Community*, www.affymetrix.com, September 2003.

[4] *CAMDA 2004 Conference, Contest Datasets*: http://www.camda.duke.edu/camda04/datasets (last access 13/06/2005).

[5] Erdal, S., Ozgur, O., Armbruster, D., Ferhatosmanoglu, H., and Ray, W.C., A time series analysis of microarray data, in: *4th IEEE International Symposium on BioInformatics and BioEngineering (BIBE 2004)*, 19–21 March 2004, Taichung, Taiwan, IEEE Computer Society, 2004, ISBN 0-7695-2173-8.

[6] Filkov, V., Skiena, S., and Zhi, J., Analysis techniques for microarray time-series data, *J. Com. Biol.*, **9**(2) (2002), 317–330.

[7] Furey, T.S., Cristianini, N., Duffy, N., Bednarski, D.W., Schummer, M., and Haussler, D., Support vector machine classification and validation of cancer tissue samples using microarray expression data, *Bioinformatics*, **16**(10) (2000), 906–914.

[8] Gardner, M.J., Hall, N., Fung, E., White, O., Berriman, M., et al., Genome sequence of the human malaria parasite *Plasmodium falciparum*, *Nature*, **419** (2002), 498–511.

[9] GenePix Pro, The image analysis software for microarrays, tissue arrays and cell arrays: http://www.axon.com (last access 13/06/2005).

[10] Griffiths, A.J.F., Gelbart, W.M., Miller, J.H., and Lewontin, R.C., *Modern Genetic Analysis*, W.H. Freeman & Co, New York, 1999.

[11] R Development Core Team, R: *A Language and Environment for Statistical Computing*, R Foundation for Statistical Computing, Vienna, Austria, 2005, http://www.R-project.org (last access 13/06/2005).

[12] Institute for Genomics Research (TIGR), www.tigr.org.

[13] Kreßel, U., Pairwise classification and support vector machine, in: B. Schölkopf, C.J.C. Burges, and A.J. Smola, Eds., *Advances in Kernel Methods–SV Learning*, MIT Press, Cambridge, MA, 1999, pp. 255–268.

[14] Lee, Y. and Lee, C.-K., Classification of multiple cancer types by multicategory support vector machines using gene expression data, *Bioinformatics*, **19**(9) (2003), 1132–1139.

[15] Le Roch, K.G., Zhou, Y., Blair, P.L., Grainger, M., Moch, J.K., Haynes, J.D., De La Vega, P., Holder, A.A., Batalov, S., Carucci, D.J., and Winzeler, E.A., Discovery of gene function by expression profiling of the malaria parasite life cycle, *Science*, **12**(301) (5639) (2003), 1503–1508. Epub 2003 Jul 31.

[16] Molla, M., Waddell, M., Page, D., and Shavlik, J., Using machine learning to design and interpret gene-expression microarrays, *AI Magazine*, **25** (2004), 23–44.

[17] Sebastiani, P., Gussoni, E., Kohane, I.S., and Ramoni, M., Statistical challenges in functional genomics (with discussion), *Statistical Science*, **18** (2003), 33–70.

[18] Simek, K., Fujarewicz, K., Swierniak, A., Kimmel, M., Jarzab, B., Wiench, M., and Rzeszowska, J.J., Using SVD and SVM methods for selection, classification, clustering and modeling of DNA microarray data, *Engineering Application of Artificial Intelligence*, **17**(4) (2004), 417–427.

[19] Suh, K.N., Kain, K.C., and Keystone, J.S., Malaria, *CMAJ*, **170**(11) (2004 May 25), 1693–1702. DOI: 10.1053/cmaj.1030418.

[20] Ward, G., Ed., Monitoring Malaria: Genomic Activity of the Parasite in Human Blood Cells, Public Library of Science, Open-access article, *PLoS Biol.*, **1**(1) (2003), 5–6.

[21] Simon, R.M., Korn, E.L., McShane, L.M., Radmacher, M.D., Wright, G.W., and Zhao, Y., *Design and Analysis of DNA Microarray Investigations*, 1st ed. Springer, 2004.

Chapter 4

Oxidative Stress Genes in *Plasmodium falciparum* as Indicated by Temporal Gene Expression

J. Noyola-Martinez[a], C. Shaw[c], B. Christian[a], G. Fox[a], M. Stevens[a,b], N. Garg[d], M.C. Gustin[b] and R. Guerra[a,*]

[a]*Department of Statistics, Rice University, Houston, TX 77005, USA*
[b]*Department of Biochemistry and Cell Biology, Rice University, Houston, TX 77005, USA*
[c]*Department of Molecular and Human Genetics, Baylor College of Medicine, Houston, TX 77030, USA*
[d]*Department of Microbiology, University of Texas Medical Branch, Galveston, TX 77555, USA*

Abstract

Entry of *Plasmodium falciparum* into human red blood cells is a stressful event for both the host and the parasite. Conversion of hemoglobin into usable food by *P. falciparum* is accompanied by the production of chemically reactive and toxic molecules called oxidants. Examination of the temporal sequence of gene expression during the intraerythrocytic development cycle (IDC) [Bozdech, Z., et al., *PLoS Biology*, 1(1) (2003), 1–16] can help elucidate how *Plasmodium* responds to these self-generated harmful chemicals while proceeding through its normal developmental program. Our study has three parts: identification of temporally-defined sets of co-regulated oxidative stress response genes in this parasite; comparison of the temporal patterns of the oxidative stress response to that of co-regulated gene sets involved in other processes; and identification of putative transcription factor binding sites by finding DNA motifs unique to the upstream regions of co-regulated oxidative stress response genes.

Keywords: malaria, oxidative stress, microarray, time-course, clustering, motif

1.   INTRODUCTION

Plasmodium falciparum is a virulent pathogen that is the major cause of the human malaria epidemic seen in developing tropical countries. *P. falciparum* utilizes mosquitoes as a vector to enter the host body and invade red blood cells. Once inside the erythrocytes, *Plasmodium* will replicate, utilizing the

---

*Corresponding author.

host's hemoglobin as a source of amino acids. After replication, the red blood cells are lysed, releasing parasites, toxic heme groups and parasitic proteins into the host's blood plasma causing the characteristic fever. The parasites will then re-enter new red blood cells and the cycle will continue, or they will be picked up by feeding mosquitoes, and will go on to infect another host.

Because of its importance as a possible Achilles heel, the *P. falciparum* oxidative stress response (OSR) has been heavily studied and many of the key proteins and enzymes have been identified [2].

After attachment and entry into the red blood cell, the parasite uses its food vacuole to engulf the concentrated hemoglobin and break the protein down into usable amino acids and leftover heme. Most of the heme forms an inert pigmented polymer inside the vacuole. However, a small amount of free heme becomes a major iron-based catalyst for formation of superoxide and other oxidants such as hydrogen peroxide. These highly reactive oxidants form co-valent bonds with proteins, nucleic acids, and lipids, thereby impairing their function. The parasite defends itself against these oxidants with enzymes that convert the oxidants into less reactive chemicals and with enzymes that re-pair the damaged cell molecules. Without the function of these enzymes, the genome and lipid membranes of *P. falciparum* are vulnerable to devastating oxidative damage. Although oxidative stress proteins have been used as a drug target for many years, the expression of the genes for these proteins has not been studied. Understanding the transcriptional activity of genes that respond to oxidative stress may be crucial to developing new drugs and gaining a better understanding of how current drugs function.

In microbes such as yeast or *E. coli*, exposure to oxidants produces a well-characterized whole genome transcriptional response [6–8,10,11]. Genes induced by oxidants include those that encode oxidant-scavenging proteins and enzymes that repair oxidant-damaged proteins, DNA, and lipid. But most of the yeast genes modulated by oxidants such as hydrogen peroxide are also regulated in the same fashion by other stressful conditions such as heat, starvation, or an increase in osmolarity caused by high sugar concentrations. This common response, termed the environmental stress response or ESR, is characterized by an increase in mRNA for stress response genes and a reduction in mRNA for genes involved in nucleic acid and protein synthesis, i.e., cell growth and division. Another striking feature of the oxidative stress response is that the increase in ESR gene mRNA is transient, falling back to control levels in less than one hour after adding hydrogen peroxide to cells. This decrease in mRNA occurs even though the stimulus, the hydrogen peroxide concentration, stays constant for a 2-hour period [8].

A potentially important difference between *Plasmodium* and other microbes concerns the timing of oxidative stress in the normal life cycle. For yeast and bacteria, a low level of oxidants is produced during normal oxygen-dependent metabolism. Immune cells and even plants attack invading microbes by producing large quantities of superoxide and hydrogen peroxide. However, the timing of this oxidative assault is highly variable and genome responses are triggered by the oxidant per se. On the other hand, for an invading *Plasmodium*, production of large amounts of oxidants (from the breakdown of hemoglobin) would occur at a fairly predictable interval after entry of the parasite. Thus, there could be two strategies in mounting an effective defense. Either the parasite is like yeast in using the oxidant as the direct inducer of oxidative stress responses of the genome. Or the protozoan would induce oxidative stress response genes as part of its developmental cycle.

From our initial analysis, we propose here a novel biological mechanism for dealing with stress. Our model is that *P. falciparum* does not initiate a general stress response during oxidative stress, but rather their stress reaction is specific and anticipatory. That is, the parasite will initiate transcription of oxidative stress response genes such as glutathione transferase and peroxiredoxin prior to the onset of that stress. This model is more probable for *P. falciparum* than it is for yeast since *Plasmodium* makes use of host hemoglobin proteins, even at the expense of having to dispose of self-generated toxic heme groups. As this metabolic process is an inherent part of the *P. falciparum* life cycle, the strategy of anticipating the accompanying oxidative stress may have a selective advantage over reacting to stress. One prediction of this model is that the increase in oxidative stress response gene expression would be co-regulated with genes involved in normal developmental processes such as making protein or RNA.

Whatever the strategy employed, the physiological situation in parasite-infected red blood cells appears very different from yeast exposed to oxidants. Oxidants cause yeast to stop making protein and RNA, to stop growing and dividing while it adapts to stress. *Plasmodium* is initiating large-scale protein and nucleic acid synthesis – using red blood cell nutrients – during oxidative stress. Expression of oxidative stress response genes continues for many hours, rather than falling as they do in yeast. This difference in timing suggests very different control mechanisms in the parasite, which may be critical to drug development.

Analytical Objective: Identifying genes co-regulated with known oxidative stress response (OSR) genes, as well as DNA sequence elements common to the upstream regions of OSR genes will be an important step forward in defining the pathways and signals that control this response. To this end, we report on a statistical analysis that identifies several promising genes and motifs that may underlie the oxidative stress mechanism in *P. falciparum*. A novel proposal is made for (1) defining selected features of time-course profiles for classification and (2) defining control groups for subsequent motif discovery.

2.   METHODS

2.1.   Study Design and Data

Bozdech et al. grew a large-scale culture of *P. falciparum* (HB3 strain) for RNA sample isolation, cDNA synthesis, labeling, and DNA hybridization with a long-oligonucleotide microarray [4]. Samples for 48 individual (hourly) time points (Cy5) corresponding to the intraerythrocytic development cycle were hybridized against a reference pool (Cy3) comprised of RNA samples representing all developmental stages (48 time points) of *P. falciparum*. In the present analysis we used the Quality Control (QC) dataset that included the set of oligonucleotides that passed all quality control filters as specified by [4] and were normalized by a linear scalar (global normalization).

2.2.   Statistical Methods

Gene profiles represented by multiple oligos were summarized by averaging the individual oligos pointwise. The QC dataset was missing time-points 23 and 29 h. Imputation for these time points, as well as for randomly missing individual points, was by averaging flanking time points. The two major analytical steps were (1) finding expression profiles similar to the OSR genes and (2) finding sequence motifs in the upstream region of all ORF's belonging to a cluster set.

Step 1: Classifying profiles by OSR genes. The objective of this step is to find genes that are potentially co-regulated with OSR genes. Having done so allows further consideration of common function and/or motifs. The approach used the entire profiles in determining similarity between OSR expression profiles and other genes. Given a reference profile (e.g., one of the OSR genes) $x = (x_1, x_2, \ldots, x_T)$ we classify all others $y = (y_1, y_2, \ldots, y_T)$ as "close" to x if the Pearson correlation between x and y is at least 0.9. Alternative distance measures, such as Euclidean distance, are possible.

A different approach is to use hierarchical clustering to let genes group naturally using Euclidean distance to agglomerate them. After the clusters were found we identified those that included the various OSR genes. These clusters can serve as the group(s) closely co-regulated with OSR genes. This second approach confirmed that genes classified by Pearson correlation are not an artifact of the reference profiles.

Step 2: Motif searching within clusters. Given gene clusters defined by the OSR genes, each is then partitioned into a training set and a testing set. The training set is generated by using a higher correlation threshold of 0.95 to our reference profile x, and the testing is composed of the genes that remain. Once the training set is established the software program MEME [1] is used to discover motifs in the training set only. MEME is not used on the testing set.

Although it's possible to find motifs in the original cluster the results would be more difficult to interpret without a valid test group. We restrict the MEME motif search to 6–8 nucleotides, since known motifs in other organisms are in this size range.

To assess the statistical significance of motifs found in the training set we developed an algorithm using R. This new algorithm interprets the MEME-motif information found in the training set and then finds the frequency of these motifs in each of the various test sets. Each motif is defined by a position-specific probability matrix (PSPM), which for each position in the sequence gives the probabilities of each possible nucleotide occurring at that position. For each motif found by MEME, our algorithm uses the PSPM to calculate P1 = *probability of the observed sequence given the PSPM (i.e. it's a motif)* and P2 = *probability of the observed sequence given the background probabilities of A, C, G, T (i.e. it's not a motif)*. We take the ratio of P1/P2 and compare it against a threshold calculated by MEME via a Bayesian method. If the ratio is greater than (less) than the threshold we accept (reject) the sequence as a matched motif. The algorithm also calculates the average motif occurrence per testing set and compares testing sets via a t-test.

Our reasoning behind all this is to verify whether motifs occurring in a given group are unique to that group. For example, given two gene clusters *c1* and *c2*, each has an associated training set *c1*-training and *c2*-training as well as a testing set *c1*-testing and *c2*-testing. MEME-motifs believed to be unique to cluster *c1* are discovered using *c1*-training and, our algorithm interrogates *c1*-testing and *c2*-testing to find if these motifs are truly unique to *c1*. If the average occurrence of motifs in *c1*-testing is greater than in *c2*-testing than it would be reasonable to propose these motifs as possible transcription binding sites unique to cluster *c1*. However, before coming to such a conclusion we must also verify that these motifs are not due to chance; therefore, we also make a comparison to a control group containing genes neither in *c1* nor *c2*. Once we determine that the motifs still occur more frequently in *c1* than in the control we are more confident about their possible candidacy for transcription control sites to any other group but *c1*. The same can be done using *c2*-training to identify possible transcription control sites unique to cluster *c2*.

2.3.   Considerations

To address the issue of whether the parasite responds or anticipates oxidative stress we considered expression patterns of genes known to be involved with hemoglobin (Hb) degradation, hemozoin sequestration/degradation and FP-binding proteins. Indeed, since Hb is the major source of oxidative stress, the expression pattern of Hb degradation-related genes in relation to antioxidant stress response genes would help confirm if the antioxidants are up-regulated in response to stress or it is an anticipated response. If the OSR genes are up-regulated before the Hb digestion related genes that would indicate anticipation of oxidative stress; up-regulation after expression of the Hb genes would signify a response to the oxidative stress.

3.   RESULTS

3.1.   Expression of Oxidative Stress Genes

P. falciparum protects itself against oxidants [2] by releasing (1) oxidant-consuming enzymes, including superoxide dismutase, peroxiredoxin, glutathione S-transferase and (2) enzymes involved in the repair of oxidized proteins and lipids reductant synthesis, including thioredoxin peroxidase, glutathione reductase, and glutathione synthetase. Figure 1 shows the time-course expression profiles for eight oxidative stress response OSR genes. The same data have also recently been used by Bozdech and Ginsburg [3] to examine the expression profiles of these and other genes in antioxidant defense in *P. falciparum*. Using the three stages of the IDC the genes may be classified according to where they (approximately peak): *Ring*: glutathione S-transferase, peroxiredoxin; *Ring/Trophozoite*: glutathione synthetase; *Trophozoite*: glutathione reductase, superoxide dismutase; *Trophozoite/Schizont*: ribonucleotide reductase, glutathione peroxidase; *Schizont*: thioredoxin peroxidase.

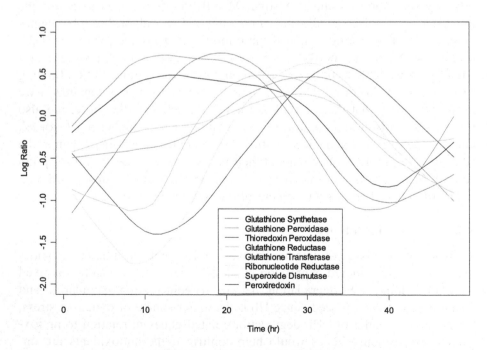

Figure 1. Oxidative stress genes.

The parasite begins significant digestion of the host hemoglobin (Hb) at the boundary between the Ring and Trophozoite stages, roughly between 12 and 18 hours into the 48-hour IDC. Therefore, relative to the digestion of Hb it may be conjectured that glutathione S-transferase and peroxiredoxin, involved in H_2O_2 dismutation, are anticipating the endogenous generation of oxidants; we call these Early OSR genes. Glutathione synthase, involved in reduced glutathione (GSH) synthesis, may be activated in response to utilization of GSH in initial response. Glutathione reductase, ribonucleotide reductase, and thioredoxin reductase react to the stress of accumulation of oxidized glutahione (GSSG) and thioredoxin ($TrxS_2$) and thus glutathione reductase may be reacting to, instead of anticipating, the oxidative stress as indicated by their sharp increase in expression coincidental with intense hemoglobin digestion. Glutathione peroxidase and catalase (both dismutate H_2O_2) may then constitute the secondary defense mechanism against oxidative stress. The remaining genes are a bit more ambiguous.

Additional information pertaining to the anticipation/reaction conjecture may be obtained by considering the expression profiles of Hb degradation-related genes. Indeed, since Hb degradation is the major source of oxidative stress the expression pattern of OSR genes in relation to those of Hb-related genes will help confirm if the antioxidants are up-regulated in response to stress or if it is an anticipated response. Here we consider vacuolar proteins which maintain the acidic environment within the food vacuole and proteases which break down hemoglobin. Figures 2 and 3 show vacuolar proteins and proteases, respectively, in relation to the Early OSR genes. The vacuolar genes are generally increasing over the Ring Stage but at a lower level than the two Early OSR genes, again supporting the hypothesis that they are anticipatory. Three of the proteases (falcipain 2 precursor, falcipain 2 precursor putative, plasmepsin 2 precursor) increase in expression over the Ring stage but the curves appear to follow the OSR genes, which also have higher levels of expression. Two proteases (plasmepsin putative, plasmepsin 1 precursor) show no change in expression and then begin to decrease at about the time that the OSR genes peak, which again seems to support a readiness by the OSR genes. All but one (falcipain 3) protease drops precipitously beginning at $\sim$18 h through $\sim$32 h. The remaining six OSR genes peak after both the vacuolar and protease genes indicating a reaction to the oxidative stress. On balance, the data support that glutathione S-transferase and peroxiredoxin anticipate oxidative stress as they peak before HB digestion and they peak or express themselves at higher levels relative to Hb-related genes.

3.2.   Classification

As described in the Methods section, genes are clustered into groups based on their tight correlation to OSR genes. Specifically four OSR genes are of

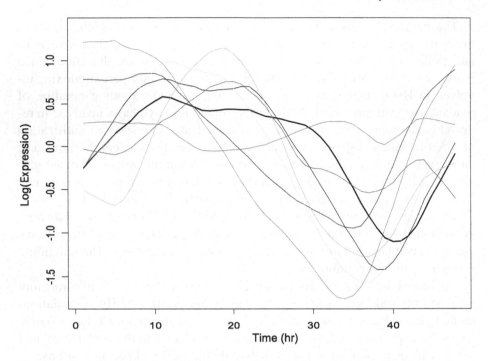

Figure 2. Protease and Early Oxidative Stress Genes. Proteases – Plasmepsin 2 precursor = red; Plasmepsin 1 precursor = green; Plasmepsin putative = dark blue; Falcipain 2 precursor = light blue; Falcipain 3 = pink; Falcipain 2 precursor putative = yellow; OSR – Glutathione Transferase = light black; Peroxiredoxin = heavy black.

interest: (1) glutathione S-transferase, peroxiredoxin which are primarily expressed early in the 48 cycle, and (2) thioredoxin peroxidase and ribonucleotide reductase which are primarily expressed late in the cycle. We propose that the two sets of Early and Late genes have different modes of transcription and that genes clustered with the Early genes have distinct transcription factors from those clustered to the Late genes. This thesis is dependent on our assumption that genes clustered to Early genes share a common transcription factor with the Early genes, and genes clustered to Late genes also share common transcription factors with the Late genes.

The Early and Late clusters, their associated training and testing groups, as well as the control set are listed below:

1. Early: $n = 157$ genes matched ($r > 0.9$) to the average profile of the two Early OSR genes (Figure 1, glutathione transferase and peroxiredoxin).
2. Late: $n = 289$ genes matched ($r > 0.9$) to the average profile of the two Late OSR genes (Figure 1, ribonucleotide reductase and thioredoxin peroxidase).

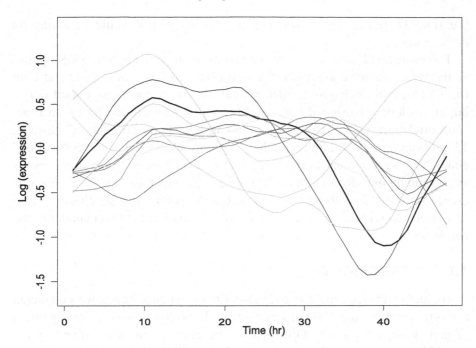

Figure 3. Vacuolar and Early Oxidative Stress Genes. Vacuolar – V. ATP Synthase = brown; V. Proton Translocation ATPase = gray; V. ATP Synthase Subunit F = purple; V. ATP Synthase Catalytic Subunit A = blue; V. ATP Synthase Subunit H = light blue; V. ATP Synthase Subunit D = red; V. ATP Synthase Subunit B = dark green; V. ATP Synthetase = light green; V. ATP Synthase Subunit E = yellow; V. ATP Synthase Subunit G = orange; OSR – Glutathione Transferase = light black; Peroxiredoxin = heavy black.

3. Early-Training: $n = 25$ genes matched ($r > 0.95$) to the average profile of the two Early OSR genes (Figure 1, glutathione transferase and peroxiredoxin).

4. Late-Training: $n = 27$ genes matched ($r > 0.97$) to the average profile of the two Late OSR genes (Figure 1, ribonucleotide reductase and thioredoxin peroxidase). We used a threshold of 0.97, instead of 0.95, to have similar sample sizes for the training sets in the Early and Late groups.

5. Early-Testing: $n = 132$ genes. These genes are obtained by taking the difference between the Early and Early-Training sets (157-25).

6. Late-Testing: $n = 262$ genes. These genes are obtained by taking the difference between the Late and Late-Training sets (289-27).

7. Control: $n = 188$ genes. These genes are mainly expressed in the middle of the 48 hour cycle so we can think of them as the Middle set.

Note that over the first 10 hours of the IDC the Early genes monotonically increase in relative expression, while the Late genes decrease in relative expression. Therefore, we surmised, that different motifs would be acting on these genes.

The Early and Late clusters listed above were obtained by using OSR genes as the reference for correlation. We also obtained Early ($n = 654$) and Late ($n = 1220$) clusters based on hierarchical clustering and these turned out to largely include the original Early and Late sets, respectively, based on Pearson correlation. Of the 157 genes found in the "Early-correlation" clusters, 131 appeared in the "Early-hierarchical" group; of the 289 "Late-correlation" genes, 274 were present in the "Late-hierarchical" set. We take these results as confirming our selection of Early and Late clusters based on the 2-gene reference sets described above. For this reason the Early and Late clusters based on correlation with the 2-gene reference sets were used in the remainder of the analysis.

3.3. Motif Discovery

To further characterize the oxidative stress response genes we conducted a motif search of the 1000bp upstream and 1000bp downstream sequences of gene groups defined by OSR genes. The Early group was defined by the $n = 157$ genes that were highly correlated ($r > 0.9$) with the average profile of the two Early OSR genes (glutathione S-transferase, peroxiredoxin). The Late group of $n = 289$ genes was similarly defined for the two Late OSR genes (thioredoxin peroxidase, ribonucleotide reductase). Since no *a priori* information concerning expected motifs was available, the two groups and a control group were used in evaluating the biological significance of a putative motif. The E-value from MEME was used as the initial screen for putative motifs and the control group was used as confirmation that a given motif was specific to a set of genes with similar expression profiles. The confirmation in the control group was carried out by our algorithm to compare motif occurrence and frequencies.

Table 1 shows the seven motifs found in the "Late-Training" group which demonstrated a significantly greater abundance in the "Late-Testing" set compared to the "Early-Testing" set. We clearly see that the average number of occurrences per motif is larger in the "Late-Testing" set. The seven motifs shown are the best examples after using a Bonferroni correction for multiple comparisons. From these results it is tempting to conclude that this method indeed selects motifs in the expected manner: MEME finds motifs in the Late-Training set, which if real should result in higher frequencies in the Late testing set than the Early testing set. Table 1 shows results supporting this expectation. However, when we used the Early-Training set to generate motifs we found no difference between Early and Late test sets. Additionally, when assessing the presence of the motifs shown in Table 1 in the control group we found that

Table 1. Motifs that statistically differentiate Early-Testing genes from Late-Testing genes based on a MEME search[1] of the 1000bp upstream regions of 27 genes correlated with Late[2] expression OSR genes

| Motif | Ave. motif occurrence per gene | | t | p-value[3] |
	Early ($n = 132$)	Late ($n = 262$)		
1 CACAT	2.08	3.37	−6.72	5.82E-09
2 ATATGTAT	1.85	3.16	−6.34	5.23E-08
3 GTGGG	1.24	1.92	−4.61	4.96E-04
4 GGGTG	1.10	1.78	−4.79	2.12E-04
5 CTTTGCA	0.80	1.51	−6.16	1.60E-07
6 GGAGTAC	0.32	0.57	−4.03	6.04E-03
7 AAAGGG	2.92	3.87	−4.04	5.95E-03

[1] Default background rates were assumed in the MEME search.
[2] See Figure 1 for Early and Late expression OSR genes.
[3] Adjusted p-values based on the Bonferroni correction are reported.

none of these motifs occurred with more frequency in the Late test group. These results were largely unexpected and made us re-evaluate our approach.

Our initial method only made use of sequences in the upstream region of the genes to find candidate motifs. In our revised approach we also considered downstream sequences. However, candidate motifs found downstream were no more abundant in either the Early or Late test sets. To try and understand this phenomenon we explored the nucleotide content in the upstream and down-stream regions of both the Late and Early sets and found the following:

| | AT/CG content (%) | | | | | |
| | Upstream | | Downstream | | Genome | |
	AT	CG	AT	CG	AT	CG
Early	88.1	11.9	84.9	15.1		
Late	86.3	13.7	84.3	15.7	80.6	19.4

The AT content in the entire genome is substantially larger than the CG content. Similar proportions are observed in the Late and Early genes in both upstream and downstream regions. Of particular interest to this study is that the AT/CG distributions are more similar between Late and Early genes in the downstream region than in the upstream region. Although the discrepancy between the Early and Late AT/CG distributions in the upstream region may seem small, it turns out that the MEME motif search is quite sensitive to such differences. Therefore, we adjusted our MEME background model to reflect the observed proportions of nucleotides. The updated results are shown in Ta-

Table 2. Motifs that statistically differentiate Early-Testing genes from Late-Testing genes based on a MEME search[1] of the 1000bp upstream regions of 25 genes correlated with Early[2] expression OSR genes

| Motif | Ave. motif occurrence per gene | | t | p-value[3] |
	Early (n = 132)	Late (n = 262)		
1 TATAATAT	8.12	4.94	7.81	3.24E-11

[1] Observed background rates were assumed in the MEME search.
[2] See Figure 1 for Early and Late expression OSR genes.
[3] Adjusted p-values based on the Bonferroni correction are reported.

Table 3. Motifs that statistically differentiate Early-Testing genes from Late-Testing genes based on a MEME search[1] of the 1000bp upstream regions of 27 genes correlated with Late[2] expression OSR genes

| Motif | Ave. motif occurrence per gene | | t | p-value[3] |
	Early (n = 132)	Late (n = 262)		
1 ACACACAT	1.20	2.66	−8.21	2.60E-13
2 TGTGTGTA	0.71	2.43	−10.42	0
3 GTGGG	1.24	1.92	−4.61	4.96E-04
4 CCTTGCG	0.08	0.26	−4.66	3.90E-04
5 GGGTG	1.10	1.78	−4.79	2.12E-04

[1] Observed background rates were assumed in the MEME search.
[2] See Figure 1 for Early and Late expression OSR genes.
[3] Adjusted p-values based on the Bonferroni correction are reported.

bles 2 and 3. Table 2 shows the motif (TATAATAT) found in the Early-Training set which is also more abundant in the Early-Testing set than the Late-Testing set. Table 3 shows five motifs found in the Late-Training set, which occur more frequently in the Late-Testing set than the Early-Testing set. Note that motifs 3 and 5 of Table 3 are also present in Table 1. Comparison of the motifs in Tables 2 and 3 to the Control group resulted in only one significant difference, reported in Table 4. Taken together, we were able to identify one motif (TATAATAT) that appears to be unique to the Early genes.

4.   DISCUSSION

When *P. falciparum* invades red blood cells it is subject to constant oxidative stress largely stemming from its digestion of host hemoglobin, as well as from reactive oxygen and nitrogen species arising from the host immune system. Using the time-course microarray data of Bozdech and Ginsburg [3] the work

Table 4. Motifs that statistically differentiate Early-Testing genes from Control-Testing genes based on a MEME search[1] of the 1000bp upstream regions of 25 genes correlated with Early[2] expression OSR genes

Motif	Ave. motif occurrence per gene		t	p-value[3]
	Early (n = 132)	Control (n = 188)		
1 TATAATAT	8.12	4.79	7.99	8.49E-12

[1] Observed background rates were assumed in the MEME search.
[2] See Figure 1 for Early and Late expression OSR genes.
[3] Adjusted p-values based on the Bonferroni correction are reported.

reported in this paper examined and characterized the gene expression profiles of eight oxidative stress response genes, the detection of other genes possibly co-regulated with these genes, and discovery of sequence motifs that may play a role in their regulation.

Our proposed model is that control of expression of oxidative stress response genes in *P. falciparum* is mediated not by an induced response to oxidants, but as a part of the normal developmental sequence important for asexual reproduction of the parasite in red blood cells. The analysis so far supports this model. The two OSR genes for peroxiredoxin and glutathione transferase increase in expression relatively early in the infection cycle, before expression of many of the enzymes needed for hemoglobin digestion. These two genes are co-expressed with genes needed for mRNA synthesis and translation. Thus, the shared function of this gene set is clearly related to the products of Hb breakdown: amino acids and oxidants. One wrinkle is that not all of the OSR genes are expressed in this Early set. For example, mRNAs for glutathione peroxidase and glutathione reductase appear later in the infection cycle. One possible explanation for different waves of OSR gene expression is that different oxidants may appear at different times. The early appearance of both peroxiredoxin and glutathione transferase mRNAs makes sense because these two enzymes are the major route for elimination of hydrogen peroxide, the most abundant oxidant appearing during Hb digestion in the Plasmodium food vacuole.

Our analysis of the upstream region of the Late OSR genes set revealed several closely related motifs (Tables 1 and 3) that are more abundant in this set than in the Early OSR genes set; for example, ACACACAT and CACAT that include CA pairs, and GTGGG, AAAGGG, GGGTG that include GGG. The biological significance remains under study, but these motifs could be binding sites for a repressor that prevents early gene expression or an activator that positively regulates gene expression at later times in the infection cycle. Our search also led to the identification of a single motif (TATAATAT) that was more abundant in the Early gene set than in *both* the Late gene set and control gene set. Our motif analysis is nevertheless preliminary, using a standard set of search parameters in only one program (MEME) to search for motifs. Additional motif analyses using MEME and other programs based on different search algorithms will be required for a more comprehensive identification of motifs or combinations of motifs that are uniquely correlated with the different OSR gene sets. The results of our study provide new insights into the biology of *P. falciparum*. We present an approach based on biological and statistical reasoning that together lead to promising areas of inquiry. Key to motif assessment in this and similar applications are (1) careful selection of classification features from the time-course profiles (e.g., time of peak expression), (2) defining appropriate control groups.

REFERENCES

[1] Bailey, T.L. and Elkan, C., Fitting a mixture model by expectation maximization to discover motifs in biopolymers, in: *Proceedings of the Second International Conference on Intelligent Systems for Molecular Biology*, AAAI Press, Menlo Park, CA, 1994, pp. 28–36.

[2] Becker, K., Tilley, L., Vennerstrom, J.L., Roberts, D., Rogerson, S., and Ginsburg, H., Oxidative stress in malaria parasite-infected erythrocytes: Host–parasite interactions, *International Journal for Parasitology*, **34** (2004), 163–189.

[3] Bozdech, Z. and Ginsburg, H., Antioxidant defense in *Plasmodium falciparum*: Data mining of the transcriptone, *Malaria Journal*, **3** (2004), 23.

[4] Bozdech, Z., Llinás, M., Pulliam, B.L., Wong, E.D., Zhu, J., and DeRisi, J.L., The transcriptome of the intraerythrocytic development cycle of *Plasmodium falciparum*, *PLoS Biology*, **1**(1) (2003), 1–16.

[5] Carmel-Harel, O. and Storz, G., Roles of the glutathione- and thioredoxin-dependent reduction systems in *Escherichia coli* and *Saccharomyces cervisiae* responses to oxidative stress, *Annu. Rev. Microbiol.*, **54** (2000), 439–461.

[6] Causton, H.C., et al., Remodeling of yeast genome expression in response to environmental changes, *Mol. Biol. Cell*, **12**(2) (2001), 323–337.

[7] Chen, D., et al., Global transcriptional responses of fission yeast to environmental stress, *Mol. Biol. Cell*, **14**(1) (2003), 214–229.

[8] Gasch, A.P., Spellman, P.T., Kao, C.M., Carmel-Harel, O., Eisen, M.B., Storz, G., Botstein, D., and Brown, P.O., Genomic expression programs in the response of yeast cells to environmental changes, *Molecular Biology of the Cell*, **11** (2000), 4241–4257.

[9] Grundy, W.N., Bailey, T.L., Elkan, C.P., and Baker, M.E., Meta-MEME: Motif-based hidden Markov models of protein families, *Computer Applications in the Biosciences*, **13**(4) (1997), 397–406.

[10] Pomposiello, P.J. and Demple, B., Global adjustment of microbial physiology during free radical stress, *Adv. Microb. Physiol.*, **46** (2002), 319–341.

[11] Zheng, M., et al., DNA microarray-mediated transcriptional profiling of the *Escherichia coli* response to hydrogen peroxide, *J. Bacteriol.*, **183**(15) (2001), 4562–4570.

Chapter 5

Identifying Stage-Specific Genes by Combining Information from Two Different Types of Oligonucleotide Arrays

Yin Liu[a], Ning Sun[b], Junfeng Liu[b], Liang Chen[c], Michael McIntosh[d], Liangbiao Zheng[b] and Hongyu Zhao[b,e]

[a]*Program of Computational Biology and Bioinformatics, Yale University, New Haven, CT 06520, USA*
[b]*Department of Epidemiology and Public Health, Yale University, New Haven, CT 06520, USA*
[c]*Department of Molecular, Cellular, Developmental Biology, Yale University, New Haven, CT 06520, USA*
[d]*Department of Internal Medicine, Yale University, New Haven, CT 06520, USA*
[e]*Department of Genetics, Yale University, New Haven, CT 06520, USA*

Abstract

The identification of stage-specific genes in the malaria parasite *Plasmodium falciparum* may provide a starting point to identify key elements for the malaria parasite to complete its life cycle. In this study, we address this question through the combined analysis of gene expression data collected from two distinct microarray platforms. Although it is intuitive that a joint analysis is likely to be more informative than that based on a single source, such analysis faces many statistical challenges in addition to the fact that different sets of genes may be probed on different platforms. First, the platforms are sufficiently different that it is difficult to correlate expression levels measured on different platforms. Second, the time resolution of the two data sets differs. To address these challenges, we have developed novel statistical methods to integrate these two distinct platforms. Based on our methods, we have identified genes that are either uniquely expressed or differentially expressed at the sporozoite and gametocyte stages. Some of these genes are known to be specific at these two stages and some are novel, providing potential candidates for transmission-blocking vaccine development. We also analyze the functions of the identified genes based on Gene Ontology (GO) classification and investigate the predicted interacting proteins. The detailed results are available at http://bioinformatics.med.yale.edu/CAMDA2004.

Keywords: microarray, sporozoite, gametocyte, nonparametric regression, gene ontology, ortholog

1.   INTRODUCTION

DNA microarray technology allows the transcription levels of many genes to be measured simultaneously, and different microarray platforms are commonly used in gene expression studies. For example, in the analysis of *Plasmodium falciparum*, the DeRisi group used microarrays based on long (70-nucleotide) oligonucleotides to quantify the relative mRNA levels of 4,488 predicted *Plasmodium falciparum* genes at 46 time points across the complete asexual intraerythrocytic developmental cycle (IDC) or asexual blood stages at a 1-hour resolution [2]. Independently, the Winzeler group employed the Affymetrix (25-nucleotide) array to examine the gene expression profiles at 6 periodic asexual blood stages, including early ring, late ring, early trophozoite, late trophozoite, early schizogony, and late schizogony stages. The parasite samples were synchronized by two independent methods: a 5% D-sorbitol treatment and a temperature cycling incubator. Besides the asexual blood stages, the gene expression levels were also measured at the gametocyte and sporozoite stages [7]. Our objective in this study is to identify genes either uniquely or differentially expressed in sporozoites and gametocytes. In our study, the genes not expressed at the asexual blood stages but expressed in sporozoites/gametocytes are defined as the genes uniquely expressed at these two stages, while the genes differentially expressed in sporozoites/gametocytes represent the genes constitutively expressed at the blood stages and up-regulated in sporozoites/gametocytes. Although the Winzeler data itself can be used alone to address this question, the higher resolution of the DeRisi data may offer additional information on gene expression during the asexual stages. Therefore, we have developed statistical methods to combine information from these two studies to fully exploit the expression data from these two different data sources. Although our methods are developed in the context of analyzing these two specific data sets, the general approach may prove useful for other similar studies in order to discover novel gene regulation patterns and to validate previous gene expression profiles. The genes identified to be uniquely or differentially expressed at the sporozoite and gametocyte stages may lead researchers to identify potential candidates for transmission-blocking vaccine development because the sporozoites are the infectious form injected to human blood by mosquitoes, and the gametocytes are the form by which the parasite is transmitted from human to mosquitoes.

2.   METHODS

2.1.   Pre-processing of the Data

For the Winzeler data, the 17 CEL files are processed using Affy R [6]. The intensity levels of the two sporozoite replicates are averaged after normalization. For the DeRisi data, in which the expression values were obtained from two-color microarray experiments with a common reference used on all the arrays, we perform the print-tip group loss normalization method within arrays by using the Limma package [5,10]. After normalization, the intensity values and log ratio values are averaged for a subset, including 8 time points that had more than one hybridization result.

2.2. Identification of Genes Uniquely Expressed at the Sporozoite/Gametocyte Stages

Our first objective is to identify genes uniquely expressed at the sporozoite/gametocyte stages, i.e. genes that are not expressed across the asexual blood stages but expressed at the sporozoite/gametocyte stages. Because the DeRisi data did not cover the sporozoite/gametocyte stages, it is not informative on its own for the identification of these genes. On the other hand, although the Winzeler data can be used to address this question, some genes that are expressed at the asexual blood stages data may be missed due to the lower resolution throughout the asexual stages in the Winzeler data. Our strategy is to first use the DeRisi data to identify genes not expressed at the asexual stages and then use the Winzeler data to examine, among this set of genes, which genes are expressed at the sporozoite/gametocyte stages. First, we need to define an objective criterion to infer whether a gene is expressed or not across the blood stages based on the DeRisi data. To achieve this goal, we utilize the 281 "EMPTY" spots on the DeRisi arrays as negative controls. For each channel, the intensities of all the "EMPTY" spots are standardized to have a mean of 0 and variance of 1 through linear transformation. The standardized intensities across all the 46 time points are then summarized. The density distributions of the standardized intensity levels for the red channel and the green channel have very similar patterns (Figure 1). Because some of the "EMPTY" spots may hybridize and yield positive signals (as suggested by the long right tails in Figure 1), we remove the spots corresponding to the upper 10% of the distribution, leaving 252 "EMPTY" spots serving as negative controls in our following analysis.

For each time point t, we calculate the mean $empMean_t$ and variance $empVar_t$ of the red channel intensities of the 252 "EMPTY" spots, and then we standardize the intensities for all other spots on the arrays by

$$R_{i,t,\text{std}} = \frac{R_{i,t} - empMean_t}{\sqrt{empVar_t}},$$

where $R_{i,t}$ represents the intensity value of spot i at time point t. The standardized intensities are summarized across all the 46 time points, so we get the

Density Plot

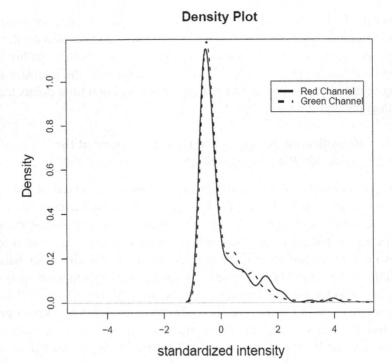

Figure 1. Density plot of the intensities in red and green channels of the "EMPTY" spots. The intensities of the 281 "EMPTY" spots have been transformed to a common distribution with mean 0 and variance 1 and summarized across all the 46 time points.

value of $R_{i,\text{std}} = \sum_{t=1}^{46} R_{i,t,\text{std}}$ for each "EMPTY" spot. The 95% percentile of these values was chosen as the expression cutoff. The genes corresponding to the spots that have summarized intensities across all the 46 time points below this cutoff are considered as genes not expressed at the blood stages.

For the Winzeler data, we need to identify genes expressed at the sporozoites/gametocyte stages. Similar to the DeRisi data, we need to choose an intensity value as cutoff to infer whether a gene is expressed or not at a specific stage. Because there are no "EMPTY" spots that we can use to derive an expression cutoff for the Winzeler data, we have to resort to other methods in our analysis. To this end, we assume that the proportion of genes not expressed at the blood stages based on the Winzeler data is the same as that based on the DeRisi data. Our previous analysis on the DeRisi data yield the result that 17% of genes are not expressed at the blood stages. Based on our assumption, we get the maximum value of the 17% percentile of gene expression levels for each of the 6 blood stages obtained from the Winzeler data and increase the value to a certain extent so that 17% of the genes can be identified as not expressed at the blood stages with taking the adjusted value as the cutoff. The genes with intensity values in the sporozoites/gametocytes above the expression cutoff are considered as genes expressed at the sporozoite/gametocyte stages. Among this set of genes, those not expressed at the blood stages are identified as genes uniquely expressed at these two stages.

2.3.   Identification of Genes Up-regulated at the Sporozoite/Gametocyte Stages

In contrast to the identification of genes uniquely expressed at the sporozoite/gametocyte stages where only a cut-off is needed to infer whether a given gene is expressed or not at a given stage, the inference of expression level changes from the combined analysis of two distinct platforms is more difficult. This requires the establishment of correspondence of measured intensity levels between the two platforms. If the two sets of data had been collected at the same time points, such analysis would be relatively straightforward if we assume that the expression level of the same gene is rather similar in the two experiments. However, the DeRisi data and the Winzeler data have rather different resolutions with 46 time points in the DeRisi data and only 6 time points in the Winzeler data across the asexual stages. To address this problem, we first identify a set of "invariant" genes, which are constitutively expressed at the asexual stages and use the measured expression levels of these genes to derive the correspondence of measured expression levels between the two datasets. For the DeRisi data, the variances of the log-ratio values $\log_2(Cy5/Cy3)$ are calculated for each expressed gene and the set of genes with a variance below a specific cutoff, 0.2 in this study, are considered as the "invariant" gene set. Similarly, the "invariant" gene set for the Winzeler data can be identified after the variances of the intensity values at the 6 blood stages are calculated. Genes common in both invariant gene sets are then selected. As the expression levels of these genes were relatively constant across the blood stages in both datasets, we calculate the mean of the gene expression values at the blood stages for each gene both based on the DeRisi data and the Winzeler data. We then apply the local linear regression method to capture the relationship between the gene expression values obtained from the DeRisi data and those obtained from the Winzeler data through $\min_{\alpha,\beta} \sum_{i=1}^{n} \{y_i - \alpha - \beta(x_i - x)\}^2 w(x_i - x; h)$.

Here, the kernel function $w(x_i - x; h)$ ensures that the observations whose covariate values x_i close to the point x are given the most weights in determining the estimate, and the smoothing parameter h controls the degrees of smoothing applied to the data [1]. The local linear estimator is

$$\hat{m}(x) = \frac{1}{n} \sum_{i=1}^{n} \frac{\{s_2(x; h) - s_1(x; h)(x_i - x)\} w(x_i - x; h) y_i}{s_2(x; h)s_0(x; h) - s_1(x; h)^2},$$

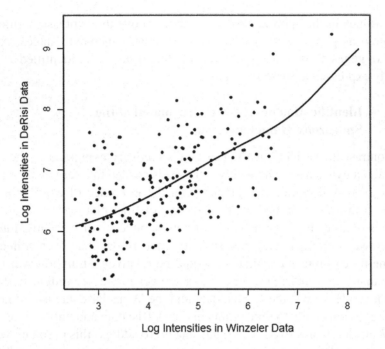

Figure 2. The nonparametric regression curve for the log intensities of all the "invariant" genes at the blood stages obtained from the DeRisi data and the Winzeler data. The smoothing parameter *h* used for control the degrees of smoothing is 1. The log intensities of "invariant" genes obtained from the Winzeler data are based on one synchronization method that uses a 5% D-sorbitol treatment.

where $s_r(x; h) = \{\sum (x_i - x)^r w(x_i - x; h)\}/n$. The results are shown in Figure 2.

Based on this nonparametric regression model, we may use the gene intensities at the sporozoite/gametocyte stages obtained from the Winzeler data to predict the values that would have been collected through the DeRisi platform. These predicted values are then compared to the measured intensities throughout the blood stages in the DeRisi data to identify genes differentially expresses at these two stages. In our study, the genes with constant expression levels at the blood stages and expression levels increased at least 1.5 fold at the sporozoite/gametocyte stages compared to the blood stages are considered as genes up-regulated at these two stages. Down-regulated genes are not considered at the two stages because we are only interested in identifying the genes directly related to the transmission between human and mosquitoes.

2.4.   Gene Ontology Analysis

Gene Ontology (GO) annotations are downloaded from PlasmoDB (http://plasmodb.org). There are 2,199 gene products (about 41% of the whole proteome) that have been assigned GO terms. We map the GO terms to the more generalized or high-level terms (GO slim terms) to gain a high-level view of gene functions. The sporozoite and gametocyte stage-specific genes are compared to the overall genes based on GO annotations using GO slim terms, and the comparisons are performed in the three GO ontologies: "molecular function", "biological process" and "cellular component". As not all the gene products were assigned a GO term, we rescale the percentages of the proteins in each GO category so that the total is 100%.

To assess the statistical significance for the GO term enrichment of the sporozoite and gametocyte stage-specific genes, we investigate whether the list of identified genes have any GO term overrepresented in their annotation compared to what would be expected by chance from the population of all the genes in *P. falciparum*. The p-value is calculated from the hypergeometric distribution as following:

$$p\text{-value} = \sum_{x}^{n} \frac{\binom{M}{x}\binom{N-M}{n-x}}{\binom{N}{n}},$$

where N represents the total number of genes in the population in which N has a particular GO term annotation. n and x represent the number of genes in the list of interest and the number of genes in the list annotated with the particular GO term, respectively. The p-value is corrected for multiple testing using Bonferroni correction, a conservative approach. There are 9 GO slim terms tested for both the "molecular function" and "biological process" terms, and 7 GO slim terms tested for "cellular component" ontology. These numbers are used for correcting the p-values. The list of sporozoite/gametocyte stage-specific genes are considered as have a GO term overrepresented compared to the overall genes if the corrected p-value is less than 0.05.

2.5.   Protein–Protein Interaction Pairs in *P. falciparum*

To study whether proteins coded by genes uniquely/differentially expressed at the sporozoite/gametocyte stages interact with each other, we utilize the interaction data from yeast because there is a lack of data for *P. falciparum*. More specifically, we perform "all-against-all" BLASTP comparisons of sequences of the *Sacchromycces cerevisiae* and *P. falciparum* proteomes, and the program INPARANOID [9] is applied on the BLASTP results to identify orthologous groups. Sequence pairs with reciprocal best hits are identified as putative ortholog pairs, and the sequences from the same species that are more similar to the putative orthologs than to any other sequences are considered as "paralogs", belonging to the same group of orthologs. Based on the concept of "interolog" [11], we assume that if protein A and protein B interact in *S. cerevisiae* and have corresponding orthologs A′ and B′ in *P. falciparum*, then A′ and B′ would form an interacting protein pair in *P. falciparum*. We use the interaction dataset for *S. cerevisiae* in the MIPS [8] database to predict interacting protein pairs in *P. falciparum* by transferring the protein interaction information between the two species.

3.   RESULTS

3.1.   Genes Uniquely or Differentially Expressed at the Sporozoite and Gametocyte Stages

The Winzeler data includes results generated from two different procedures to synchronize *P. falciparum*. We identify sporozoite/gametocyte stage-specific genes using data generated from both synchronization procedures. Table 1 summarizes the results of our study. As shown in Table 1, both synchronizations yield similar results with an almost complete overlap between different synchronizations.

A total of 408 genes are found to be expressed at the sporozoite stage but not expressed at the asexual blood stages, and 118 genes are constitutively expressed at the asexual blood stages and up-regulated at the sporozoite stage. Among these genes, some of them are experimentally known to be sporozoite specific. For example, the sporozoite surface protein 2 and the circumsporozoite surface protein are well-known markers of the sporozoite stage and are included in our identified gene set.

Similarly, a total of 124 genes constitutively expressed at the asexual blood stages are up-regulated at the gametocyte stage. An additional set of 335 genes is identified as expressed at the gametocyte stage but not at the asexual stages. Included in this list are well-known gametocyte-specific genes, such as those encoding meiotic recombination protein DMC1 and 25kDa ookinate surface antigen. Compared with the results in the Winzeler study, our gene set in-

Table 1. The number of sporozoite and gametocyte stage-specific genes. In the category of "Constitutively expressed", the genes up-regulated at the sporozoites/gametocyte stages are listed. In the category of "Not expressed", the genes uniquely expressed at the sporozoite/gametocyte stages are listed

Expression pattern at the asexual blood stage	Sporozoite			Gametocyte		
	Sync1	Sync2	Overlap	Sync1	Sync2	Overlap
Constitutively expressed	120	139	118	124	140	124
Not expressed	418	411	408	346	339	335
Total	538	550	526	470	479	459

cludes 76% of 61 genes identified as sporozoite specific and 69% of 210 genes identified as gametocyte specific in the Winzeler study, respectively.

Besides genes that are known to be stage-specific, we have also identified some genes that have not previously been shown as sporozoite- or gametocyte-specific in the Winzeler study. For example, the protein encoded by MAL13P1.304 is a potential malaria surface antigen and was identified as up-regulated at the sporozoite stage in our results. In addition, MAL6P1.195, encoding a RNA-binding protein MEI2, has been found to be specifically expressed in gametocytes in our analysis. Although the proteins encoded by these genes have been identified as sporozoite/gametocyte specific in the proteomics study based on mass spectrometry data [3], these genes were not identified as sporozoite- or gametocyte-stage specific in the Winzeler study. Therefore, our methods may provide a more comprehensive list of stage-specific genes that are worthy of further investigation and may represent potential candidate targets for the development of transmission-blocking vaccines.

3.2. Gene Ontology Classification

The comparisons of GO annotations with high-level GO terms between the sporozoite/gametocyte stage-specific genes and the overall genes are shown in Figures 3a–c, and the list of GO terms associated with a significant p-value are provided in Table 2.

In the "molecular function" ontology, a higher percentage of proteins encoded by the sporozoite/gametocyte uniquely expressed genes are assigned to

Table 2. The list of GO terms overrepresented by sporozoite and gametocyte stage-specific genes. The p-values are calculated from hypergeometric distributions and corrected for multiple testing using Bonferroni correction. The GO terms associated with a corrected p-value less than 0.05 along with the corresponding gene set are listed. The full list of GO terms associated with their p-values is available online

GO term		Gene set	Corrected p-values
Molecular function	Defense/immunity protein	Sporozoite expressed	1.24E-11
		Gametocyte expressed	1.40E-9
	Cell adhesion	Sporozoite expressed	5.91E-12
		Gametocyte expressed	5.75E-10
Biological process	Cell communication	Sporozoite expressed	2.60E-13
		Gametocyte expressed	5.05E-7
	Cell adhesion	Sporozoite expressed	5.91E-12
		Gametocyte expressed	5.75E-10
Cellular component	Extracellular	Sporozoite expressed	9.46E-13

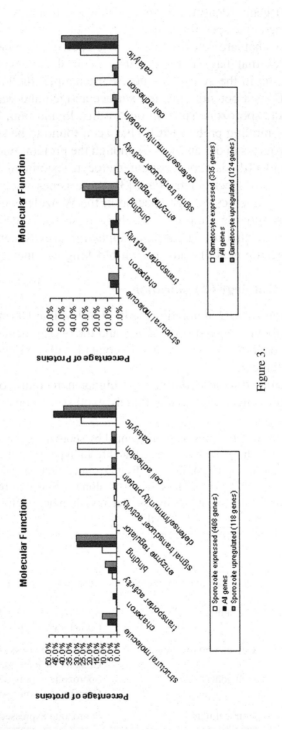

Figure 3.

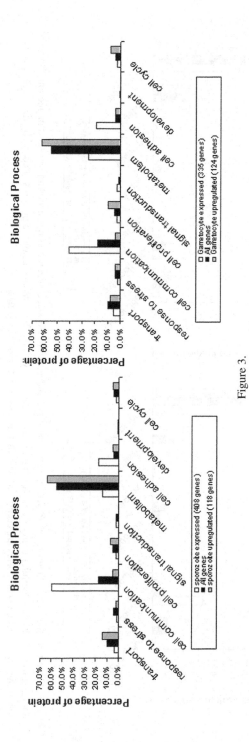

Figure 3.

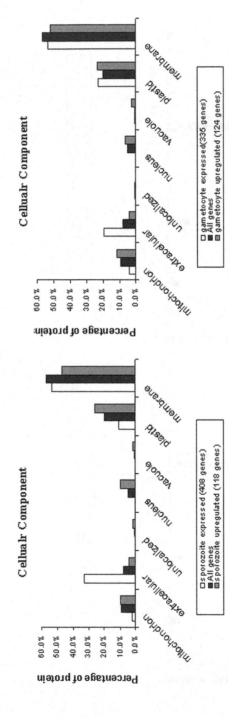

Figure 3. Gene Ontology classifications of *P. falciparum* sporozoite and gametocyte stage-specific genes according to the "molecular function" (a), "biological process" (b) and "cellular component" (c) ontologies of the GO system. The percentages of the proteins encoded by the stage-specific genes in each of the high-level GO categories are compared with that of all the *P. falciparum* genes.

the "defense/immunity protein" and "cell adhesion" categories compared to the overall gene products. And the statistical analysis provides the evidence that the sporozoite/gametocyte uniquely expressed genes have these two GO terms overrepresented (Table 2). This result is reasonable as the genes specific in sporozoites/gametocytes are involved in the evasion of the host immune system and the cell communication process.

Among all the categories in the "biological process" ontology, the identified stage-specific genes, including 34% of sporozoite specifically expressed genes and 24% of gametocyte specifically expressed genes are over-represented in the "cell communication" category with p-values of 2.60E-13 and 5.05E-7, respectively. These cell communication related genes are known to be involved in "host-pathogen interactions" or "cell–cell adhesion" processes, which may reflect the specific processes relevant to the sporozoite and gametocyte stages [4].

In the "cellular component" ontology, a higher percentage of sporozoite specific gene products belong to the "extracellular" category (with p-value of 9.46E-13). More detailed analyses reveal that this is mainly due to the large number of erythrocyte membrane protein 1 and rifin genes in our identified gene set, and these genes have been shown as sporozoite/gametocyte specific in previous studies [3,7].

We also compare the GO enrichment of our identified genes with the results from the Winzeler study. We select the genes identified as gametocyte specific in the Winzeler results but are not included in our identified gene set and perform GO analysis on these genes (Figure 4). According to the "molecular function" and "biological process" ontologies, these genes do not show different GO term enrichment compared to the overall gene products (with p-values larger than 0.05, supplementary data online). This suggests that these genes as a group are different from the genes identified from our set.

3.3. Correlate Protein Interaction with Gene Expression

Based on comparative study, only 935 *P. falciparum* proteins have corresponding *S. cerevisiae* orthologs, and a total of 646 interacting protein pairs among these 935 proteins are predicted based on the ortholog list. There may be correlation between expression patterns among the interacting protein partners because the functionality of the interacting pairs depends on the presence of two proteins participating the interaction. To test our hypothesis, we study the number of interacting protein pairs among the sporozoite and gametocyte stage-specific genes and the results are summarized in Table 3.

Because there are 15 proteins having more than 5 interacting partners, we evaluate the statistical significance of the observed number of interacting pairs through simulations after removing these so-called "hub" proteins. Specifically, we randomly select the same number of proteins from the ortholog list

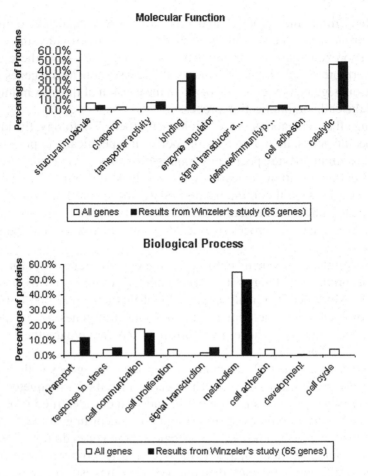

Figure 4. Comparisons between the genes identified as gametocyte specific in the Winzeler results but not included in our identified gene set and all genes according to the "molecular function" and "biological process" ontologies of the GO system.

Table 3. Interacting protein pairs in sporozoites and gametocytes. The empirical statistical significance is calculated as the fraction of the 10,000 permutations having a larger number of protein pairs than that based on the observed data

	Number of proteins having yeast orthologs	Number of protein pairs within the gene set	Empirical statistical significance
Sporozoites	62	5	0.0405
Gametocytes	54	5	0.0396
Whole Proteome	935	646	

(e.g., 62) and record the number of interactions among these randomly selected proteins. This procedure is repeated 10,000 times and the statistical significance of the observed number of interacting pairs can be estimated based on the 10,000 simulated results. As shown in Table 2, there is marginal evidence suggesting that the proteins in the list are more likely to interact with each other than expected by chance.

Only a small number of genes with orthologs in *S. cerevisiae* are found to be sporozoite/gametocyte stage-specific, resulting in the identification of only a few protein–protein interactions at these two stages. Of the 5 interacting protein pairs found in gametocytes or sporozoites, 4 were found to be common at both gametocyte and sporozoite stages (MAL7P1.50 and PF07_0139, PF00_0002 and MAL7P1.145, PF10_0258 and PF11_0481, PF10_0258 and PF14_0030). Among them, both PF00_0002 and MAL7P1.145 play a role in DNA mismatch repair, an important process for *P. falciparum* reproduction at the gametocyte stages and perhaps required in sporozoites in preparation for extensive DNA replication and schizogony, which occurs following the invasion of hepatocytes.

4.   CONCLUSIONS AND DISCUSSION

The identification of stage-specific genes provides a starting point to identify key regulatory elements essential for the malaria parasite to complete its life cycle. In this article, we have developed statistical methods to combine information from two datasets generated under different microarray platforms to identify genes either uniquely expressed or differentially expressed at the sporozoite/gametocyte stages compared to the asexual stages. Our identified genes show significant enrichment for certain Gene Ontology categories related to the functions and processes involved in the sporozoite/gametocyte stages. Although the genes identified in our study have a high degree of overlap with those from the Winzeler study, we did not observe any functional enrichment for those genes identified in the Winzeler study but not in our analysis, suggesting that our methods have a higher degree of specificity. By combining information for two different sources, we were able to take advantage the higher resolution of the DeRisi data (as compared to the Winzeler data) to study gene expression patterns at the two stages that were only collected in the latter study. This combined analysis allowed us to identify a larger number of genes that are up-regulated at the gametocyte and sporozoite stages than that based on one data source where the time resolution is low. It is conceivable that even more information can be extracted from other data sources, if they become available, to better understand the mechanisms responsible for the transmission of this protozoan malaria.

Only a small number of genes with orthologs in *S. cerevisiae* were found to have different expression patterns in gametocytes and sporozoites, resulting in the identifications of only a small number of protein–protein interactions. Our simulation results indicated marginal evidence of increased likelihood of interactions among stage specific proteins. These interacting proteins may serve as effective targets for blocking transmission by anti-malaria drug or vaccine development, as they are likely to be involved in both sexual stage development as well as invasion of the human host.

ACKNOWLEDGEMENTS

We thank Nianjun Liu and Naixin Li for help discussion regarding this study. This work was supported in part by NSF grant DMS-0241160.

REFERENCES

[1] Bowman, A. and Azzalini, A., *Applied Smoothing Techniques for Data Analysis*, Clarendon Press, Oxford, 1997.

[2] Bozdech, Z., Llinas, M., Pulliam, B.L., Wong, E.D., Zhu, J., and DeRisi, J.L., The transcriptome of the intraerythrocytic developmental cycle of Plasmodium falciparum, *PLoS Biol.*, **1**(1) (2003), E5.

[3] Florens, L., et al., A proteomic view of the *Plasmodium falciparum* life cycle, *Nature*, **419** (2002), 520–526.

[4] Gardner, M., et al., Genome sequence of the human malaria parasite *Plasmodium falciparum*, *Nature*, **419** (2002), 498–511.

[5] http://bioinf.wehi.edu.au/limma

[6] http://biosun01.biostat.jhsph.edu/Eririzarr/Raffy/

[7] Le Roch, K.G., Zhou, Y., Blair, P.L., Grainger, M., Moch, J.K., Haynes, J.D., De La Vega, P., Holder, A.A., Batalov, S., Carucci, D.J., and Winzeler, E.A., Discovery of gene function by expression profiling of the malaria parasite life cycle, *Science*, **301**(5639) (2003), 1503–1508.

[8] Mewes, H.W., et al., MIPS: Analysis and annotation of proteins from whole genomes, *Nucleic Acids Research*, 32 Database issue: D41-4, 2004.

[9] Remm, M., Storm, C.E., and Sonnhammer, E.L., Automatic clustering of orthologs and in-paralogs from pairwise species comparisons, *J. Mol. Biol.*, **314**(5) (2001), 1041–1052.

[10] Yang, Y.H., Dudoit, S., Luu, P., and Speed, T.P., Normalization for cDNA microarray data, *SPIE BiOS* (2001).

[11] Yu, H., Luscombe, N.M., Lu, H.X., Zhu, X., Xia, Y., Han, J.D., Bertin, N., Chung, S., Vidal, M., and Gerstein, M., Annotation transfer between genomes: Protein–protein interologs and protein-DNA regulogs, *Genome Res.*, **14**(6) (2004), 1107–1118.

Chapter 6

Construction of Malaria Gene Expression Network Using Partial Correlations

Raya Khanin and Ernst Wit

Department of Statistics, University of Glasgow, Glasgow, UK
e-mail: raya@stats.gla.ac.uk

Abstract
In this paper we model the gene expression network of *Plasmodium falciparum* using the time-course microarray dataset [Bozdech, Z., et al., *PLoS Biol.*, **1**(1) (2003), E5] A gene expression network is constructed based on a novel method that combines two types of correlations between each pair of genes: standard Pearson and partial correlations. A link is established between two genes if both correlation coefficients are higher than their corresponding thresholds. The values for thresholds are sought so that the topology of the resulting network satisfies several criteria. The sought network has to be sparse, small-world (with any two genes being connected by a path of a few links only), scale-free-like (wherein a small number of genes have a large number of links and many genes have only a few connections). Similar to gene networks of other organisms the highly connected genes (hubs) in the constructed network tend to have essential cell functions. To verify the proposed method and to compare the results, a scale-free-like, small-world gene expression network was also constructed using another dataset [Le Roch, K.G., et al., *Science*, **301**(5639) (2003), 1503–1508], confirming the lethality and centrality property of malaria hubs.

Keywords:
gene expression network, partial correlation, scale-free-like network

1. INTRODUCTION

The objective of this study is to construct a gene expression network of *Plasmodium falciparum* using the time-course microarray data-set from Bozdech et al. [3]. Unravelling the topology of the malaria gene network is relevant to understanding cell function and the invasion cycle of the parasite. We use a graph-theoretical approach where nodes in the network stand for genes and edges between two nodes stand for links representing relationships or associations between the two genes. In the network, the genes (nodes) are connected if certain criteria, such as co-expression, are satisfied.

Analyses of gene co-expression networks have shown a correlation between the essentiality of a gene and the number of connections that the gene has: highly connected genes (hubs) are often essential (involved in central biological functions) and evolutionarily conserved [2,16]. For *Plasmodium falciparum* more than 60% of predicted 5409 open reading frames lack sequence similarity to genes from any other known organism [3]. In addition, 65% of all annotated genes encode hypothetical proteins of unknown functions. This makes ascribing putative roles for such genes a challenging task. One of the potential benefits of gene network analysis is to obtain clues on the putative roles of such genes of unknown function based on the gene connectivities, positions in the network, and the other genes with which they have links.

It is of some interest to see whether the gene network analysis can give some support to the hypothesis advanced in [3] on a regulatory network wherein a comparatively small number of transcription factors with overlapping binding site specificities could account for the entire cascade. The authors speculated further that disruption of a key regulatory element (lethal gene) might have a profound inhibitory effect on the entire network [3]. Such lethal genes are most likely to be among the highly connected nodes in the malaria network.

For the study of the malaria gene regulatory network, we used two datasets. The first is the *overview* dataset from the complete intra-erythrocytic developmental cycle (IDC) transcriptome of *Plasmodium falciparum* measured at 46 time-points [3]. To verify results, we have also used a time-course dataset measured at nine time-points in human and mosquito stages of malaria parasite's life-cycle [10]. We will further refer to this dataset as the *validation* dataset.

2.   NETWORK CONSTRUCTION FROM TOPOLOGICAL CONSTRAINTS

We aim to construct a network of malaria gene interactions, using global topology constraints, which have been found to be characteristic for other biological networks. These constraints include network sparseness, the small-world property, and the existence of a few highly connected nodes and many genes with a few connections.

An important measure of networks topology is the distribution of the number of connections per node. The number of connections per node is often called the *connectivity* of a node or its *degree*. Therefore, the distribution is referred to as the *connectivity (or degree) distribution*. Previously studied biological networks of interactions, including gene expression networks of other organisms, have shown to have many nodes with few connections and a few nodes with many connections (hubs) [1,2,11,16]. The existence of hubs has often been cited as the most characteristic feature of biological networks and in particular of the scale-free networks [1,2,16].

Although, the networks are commonly referred as being scale-free, it is their connectivity distribution that is considered to be scale-free. Precisely, distribution is defined as scale-free if its relative frequency distribution is given by a power-law, $p(k) \sim k^{-\gamma}$, $k \geqslant 1$, where k stands for the connectivity of a node and γ is the power-law exponent. It has recently been reported that the evidence collected to support the scale-free property of biological networks is questionable [15]. It has also been found that the connectivity distribution in many inferred biological networks differs in a statistically significant way from the power-law, and these networks are, strictly speaking, not scale-free [9]. In addition, a plausible evolutionary mechanism such as evolutionary drift is not compatible with scale-free distribution [13]. However, certain characteristics of a scale-free network, such as a small-world property and the existence of hubs, are valid for real genetic networks, and in the absence of consensus on an alternative distribution, the power-law can be used for modelling purposes as a first-order approximation. In particular, the connectivity distribution described by a power-law can be useful for construction of global gene regulatory networks, whose structure is mainly unknown. In this paper, we will be looking for the network connectivity distribution that resembles a power-law. We will refer to such networks as *scale-free-like*.

A chi-squared statistic $T = \sum_{k=1}^{k^*}(O_k - E_k)^2/E_k \sim \chi^2_{k^*-2}$ has been used as a measure of closeness of a network's connectivity distribution to scale-free behaviour. Here O_k are the observed (constructed) values of connectivities from the data, and E_k are the values estimated from the power-law, with γ estimated by the maximum likelihood method as described below. The connectivity values over k^*, for which the expected number of connections is less than 5, are pooled together. The smaller the value of the chi-squared statistic T the closer the connectivity distribution resembles a power-law.

For several gene co-expression networks, whose connectivity distribution has been modelled by the power-law, the power exponent γ has been reported to be of the order of 1.0 [2,16]. We have determined the power-exponent, $\hat{\gamma}$, of the network under consideration, by the maximum likelihood method from fitting the power-law distribution $p(k) = \frac{k^{-\gamma}}{\zeta(\gamma)}$ to the constructed connectivities (or degrees). Here $\zeta(\gamma)$ is the (truncated) Riemann zeta-function and $k \geqslant 1$. The number of connections (connectivity), x_i, for a node i is often obtained from experimental or simulated data.

In a large network the number of connections of different nodes can be assumed to be approximately independent. We have shown elsewhere [9] that the assumption of independence of connectivities of all nodes in the network can be weakened by assuming independence of connectivities of nodes in a smaller sub-network. As a result, the likelihood function can be written as $L(\gamma|x) = \prod_{i=1}^{N} x_i^{-\gamma}/\zeta(\gamma)$, where N is the maximum connectivity. The log-likelihood $l(\gamma|x) = -\gamma \sum_{i=1}^{N} \log x_i - N \log \zeta(\gamma)$ is maximized by finding zeros of its derivative using the standard Newton–Raphson method for finding roots of the function.

Another important property gene expression networks have been shown to possess is a *small-world* property. In simple terms, this property implies that any two nodes can be connected with a path of only a few links. The small-world property is often quantitatively characterized by a large average clustering coefficient, C, which reflects the connectedness of the neighbours of a given node between themselves. The clustering coefficient of a gene i is computed by $c_i = 2n_i/k_i(k_i - 1)$, where n_i is the number of links connecting the k_i neighbours of gene i forming triangles and $k_i(k_i - 1)/2$ is the total number of triangles that could pass through the node i. The average clustering coefficient, C, of small-world networks is typically several orders of magnitude higher than that of a random network of equivalent average connectivity and size $C_r \approx k/N_{\text{genes}}$.

In addition, gene regulatory networks are known to be *sparse* because genes influence and/or are being influenced by a limited number of other genes [1]. This implies that average number of connections (connectivity) per gene (node), k is not large. Theoretical studies found the values for average connectivity in gene expression networks of different organisms to be of the order of 10–30 [11,16]. In this work, we will be looking for a network with the average connectivity in this range.

3.   METHOD FOR THE CONSTRUCTION OF EXPRESSION NETWORK

The main thrust of this paper is to construct a malaria gene expression network based on thresholding pairwise Pearson correlations and partial correlations of gene profiles.

The threshold parameters were sought so that the constructed network satisfies four global topological criteria, described above. (1) The network is sparse, with an average connectivity, k, of the order of magnitude of 10; (2) the network has the small-world property such that is characterized by a clustering coefficient which is much higher than that of a random network with the same average connectivity and size, $C_r = 10/3000 = 0.003$; (3) the connectivity distribution is scale-free-like, i.e. it is as close as possible to the power-law, as seen in yeast and other organisms [2,4,11,16]; and (4) the power-law exponent $\hat{\gamma}$ of the connectivity distribution is close to 1.0 as has been reported for other gene expression networks [2,11,16].

3.1.   Pearson Correlation

There have been a number of studies where global gene networks are constructed from microarray data based on the Pearson correlation coefficients. Two genes are considered linked in the co-expression network if their correlation is higher than the threshold [2,16]. Sometimes one also takes into account empirically calculated p-values for the correlations between two genes [4]. The Pearson correlation has been shown to play an important role in inferring interactions between genes [7]. However, methods that are based only on standard correlations are too simplistic and inevitably overestimate the number of links (connectivity) per gene. It is common knowledge that a high correlation coefficient is indicative not only of nodes that have direct connections but also of nodes with indirect connections. It is also plausible that some important true connections are left out if the threshold is not low enough. However, lowering the correlation threshold will significantly increase the number of potential links, including many random ones.

In the case of the malaria time-course dataset, the problem of including too many random links becomes even more transparent due to a very highly co-ordinated expression of genes [3]. A network constructed from the overview malaria dataset by thresholding correlations, while restricting the average connectivity per node, k, results in very high threshold values, R. For example, to obtain a network with $k = 50$ the threshold $R = 0.935$ is required. Restricting the average connectivity to a lower value, $k = 30$, results in an even higher value of threshold, $R = 0.95$. This is an unreasonably high value. Given the noisy data, missing values and the complexity of biological networks, many biologically relevant connections will not be included in such network.

For a slightly lower value of Pearson correlation cut-off, $R = 0.8$, the constructed network ceases to be sparse. In addition, its connectivity distribution is not scale-free-like (Figure 1). In fact, this co-expression network includes about 15% of all possible links, with an average number of links per node, $k = 470$, being more than ten times higher than the average connectivity for the gene networks of other organisms constructed by the same method. For example, with $k = 32$, the sparse scale-free network of yeast was constructed with only $R = 0.6$ [16].

3.2.   Partial Correlation

Here we propose to use partial correlations to filter the more likely links out of a much larger set of potential links with high standard correlations. The partial correlation coefficient of two genes measures the strength of relation between these genes after the effect of other genes is removed or fixed, therefore indicating whether two genes are directly or indirectly linked. The partial correlations of different orders have been used in Gaussian Graphical

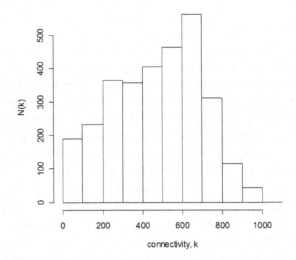

Figure 1. Histogram of connectivities in a malaria co-expression network constructed with a threshold $P = 0.8$ from overview dataset [3]. The average connectivity per node is $k = 470$ and the network is not scale-free. There are several highly connected genes and a much larger number of genes links with connectivities in the medium range.

Models (GGM) to characterize strength of correlations between pairs of genes in regulatory networks [12,14,17]. First-order partial correlations have been used to elucidate the regulatory network of *Arabidopsis thaliana* [17] and *Sacchromycces cerevisiae* [12]. These authors consider all possible triangles of three genes to explore the dependence between two of the genes conditioned on the third. All these triangles are then combined to make inferences on the complete network using either frequentist or latent random graph approaches. Second-order partial correlations, conditioning each pair of genes on every other pair of genes, have been applied to computer simulated networks and to yeast gene expression data [5]. Another method uses full-order partial correlations (conditioned on all other genes in the network) and the false discovery rate (FDR) approach to infer edges of the gene network from both simulated and real microarray data [14].

We propose to construct a gene expression network from a large gene dataset by using both Pearson and (full-order) partial correlation coefficients for each pair of genes. Namely, for each pair of genes (i, j) we compute the Pearson correlation of their profiles, r_{ij}, and their partial correlation coefficient, q_{ij}. The partial correlation of genes i and j with respect to other genes whose effect is removed (fixed) is given by

$$q_{ij} = \frac{\omega_{ij}}{\sqrt{\omega_{ii}\omega_{jj}}},$$

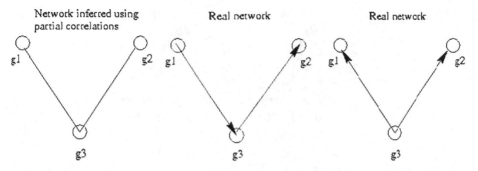

Figure 2. Schematic figure of the drawbacks of a representation of gene regulatory relationships by an undirected network. If in the inferred network, gene g3 is connected to genes g1 and g2 by undirected links (left), then it is impossible to distinguish between several scenarios in the real network. For example, gene g1 regulates gene g3, which in turn regulates gene g2 (middle), or gene g3 regulates genes g1 and g2 (right). Two other variants are possible.

where $\omega_{ij} = \{r_{ij}\}^{-1}$ is the inverse of the Pearson correlation matrix, $\{r_{ij}\}$. To overcome the degeneracy problem of the correlation matrix $\{r_{ij}\}$ for small samples, partial correlation estimators based on the Moore–Penrose pseudo-inverse of correlation matrix were introduced in [14]. In our work we follow this approach and compute partial correlations by using the Moore–Penrose pseudo-inverse of the correlation matrix via the *cor2pcor()* function from *R*-package *GeneTS* [14]. Two genes (i, j) are connected by a link if their Pearson correlation is higher than a cut-off value, R, and their partial correlation is higher than (or equal to) a cut-off value, Q:

$$i \leftrightarrow j: r_{ij} \geqslant R \text{ and } q_{ij} \geqslant Q.$$

The general drawback of any inference approach that results in an undirected network (such as a GGM) is that it gives no indication of causality. A link connecting two genes does not indicate which gene in the pair is the regulator and which is the regulated one, as illustrated in Figure 2. Although lacking causality information, undirected networks are a very useful first level representation of gene regulatory relationships on a genome wide level. Further levels of representations are directed networks, where the direction of the regulatory relationship is specified. This can eventually be extended by quantitative information, such as probabilities of connection in Bayesian networks or kinetic parameters of regulation.

4. RESULTS

For the overview dataset, the values from multiple oligonucleotides representing the same gene were averaged, resulting in expression values for 3048

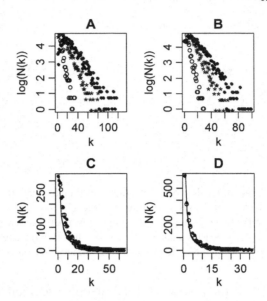

Figure 3. Connectivity distribution of nodes in a malaria gene network constructed from the overview dataset for different values of thresholds. (A), (B) Log distribution of connectivities for $r = 0.45; 0.5; 0.55$ and $P = 0.7$ (A) and $P = 0.8$ (B). (C), (D) Distribution of observed connectivities and fitted power-law $N(k) \sim k^{-\gamma}$ for $r = 0.5$ and $P = 0.8$, $\hat{\gamma} = 0.91$ (C) and $P = 0.7$, $\hat{\gamma} = 0.84$ (D).

genes. In the rest of the paper we will concentrate on reporting the results for the overview dataset. The topology of the network constructed using the validation dataset is very similar (see Tables S5 and S6 on the supplemental web-page: www.stats.gla.ac.uk/~raya/Malaria/suppldata.html).

We have performed a grid search for the threshold values R and Q based on topological criteria. We have found a range $0.45 \leqslant Q \leqslant 0.6$ and $0.7 \leqslant R \leqslant 0.8$ for which all four topological constraints are satisfied. The qualitative topological properties of the malaria network are insensitive to the precise thresholds within this range of values. Taking the thresholds within this range yields a scale-free-like distribution of connectivities, which are qualitatively similar. Figure 3 shows connectivity distributions, $N(k)$, for several values of thresholds R and Q. Values outside this region result in other types of networks. $Q \leqslant 0.4$ results in networks whose connectivities do not obey a power-law (Figure 4); while $Q > 0.6$ and/or $R > 0.8$ yield too few links (not shown).

Values of $\hat{\gamma}$ are within the range 0.6–1.4 for different values of thresholds Q and R. $\gamma = 0.6$ for the parameters $R = 0.7$, $Q = 0.45$ produce a network with an average connectivity per node of $k = 28$ and maximum connectivity $k_{\max} = 133$, and $\hat{\gamma} = 1.4$ is for $R = 0.8$, $Q = 0.6$ with $k = 4$, $k_{\max} = 30$. Other values of parameters resulted in networks with average connectivities between these two values (see Table S1 on the supplemental web-page).

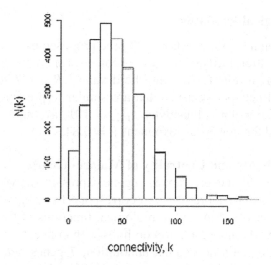

Figure 4. Histogram of connectivities in a malaria co-expression network constructed with thresholds $P = 0.7$, $r = 0.4$ from overview dataset [3]. Lowering one of the thresholds outside the accepted region results in a network whose behaviour is very different from scale-free.

The clustering coefficients have been found to be within the range $C = 0.195$ for $R = 0.7$, $Q = 0.45$ and $C = 0.443$ for $R = 0.8$, $Q = 0.6$. These values are much higher than the value for random networks of equivalent average connectivity and size ($C = 0.003$), and they are consistent with the values reported for other organisms (e.g., $C = 0.6$ for yeast [16] and other organisms [2]).

4.1.   Statistical Validation

To find whether a network constructed by thresholding the two types of correlation coefficients is statistically meaningful or whether it can easily be found by chance, we performed a permutation test. For each gene we reshuffle the values at each time-point, constructing a gene profile of the same length, with the same values but with a different time-order of these values. We then recompute the correlation and partial correlation matrices and establish a link between genes i and j if the thresholding conditions ($R = 0.7$, $Q = 0.5$) are satisfied. In 100 permutation networks, two links are found on average for each network (estimated standard error $= 0.14$) compared to several thousands in the network inferred from the original dataset. This allows us to conclude that the network inferred by the thresholding method is unlikely to have arisen by chance.

4.2.   Biological Validation

The expression network constructed by the proposed method in Section 4.1
is worth investigating further for some proof-of-principle results. In the next
section we report results for the threshold values $R = 0.7$, $Q = 0.5$. These
parameters yield network statistics that are similar to previously studied networks with a maximum connectivity $k_{\max} = 101$, average connectivity per
node $k = 15$, and the power-law exponent $\hat{\gamma} = 0.84$.

4.2.1.   Lethality and Centrality of Malaria Genes.   It has been previously reported that high degree nodes in gene expression networks constructed
for other organisms are more likely to correspond to essential and conserved
genes, i.e. to be involved in central biological functions of the cell [2]. In the
constructed network, among the top 66 hubs with connectivities from $k_{\max}/2$
there are 13 genes with no manual annotation, 7 genes belong to the Plastid genome, and 30 genes code for proteins with unknown functions (hypothetical proteins). Therefore, only 16 hub-genes code for proteins with some
identifiable functions. Among them, 7 genes (PFI1340w, PFI1360c, PFI0385c,
PF13_0229, PF14_0373, PFA0345w, PF11_0298) are known to have essential
functions in cell growth, maintenance, and metabolism (according to GO annotation). In addition, a rhoptry protein (PFI0265c), a papain family cysteine
protease (PFI0135c), and an early transcribed membrane protein (PF10_0019)
are also in the list of the hub-genes. Among 5 hubs on chromosome 9, three
(PFI1340w, PFI1360c, and PFI0385c) are prescribed functions in cell growth,
maintenance and metabolism, and they are all connected among themselves
forming a triangular network motif. The largest reported ORF (MAL6P1.147)
also has a large number of links, half of maximum connectivity. Other 8 annotated hubs out of 30 that code for hypothetical proteins are either conserved or
have homologues/similar to proteins in other organisms.

The list of 66 top hubs for the network constructed from the validation
dataset with $R = 0.8$, $Q = 0.5$ contains 20 genes (virtually all annotated hubs)
with cell growth/maintenance, cell communication, and other central cell functions. For a full list of hubs in networks constructed for the overview and the
validation datasets see Tables S2 and S6 on the supplemental webpage.

As another proof-of-principle, we looked at how many hubs are in the set
of only 6% of all genes in the genome of *Plasmodium falciparum* that were
found to be common to all four stages of the parasite life cycle (supplementary Table 1 in [6]). This list contains primarily housekeeping genes and their
products, such as ribosomal proteins, transcription factors, and cytoskeletal
proteins. It turned out that 15 hubs from our list are among this set of common
genes found in [6]. This is about 30% of all hubs with manual annotation.

It is of interest to see whether genes with unknown functionality among the
hubs belong to classes of essential genes. We looked at how hubs that code

for protein with unknown functions in the overview network clustered in the experiments of Le Roch et al. [10], as it has been demonstrated for various organisms that genes that cluster together are more likely to have similar biological functions. We found that among 25 genes coding for hypothetical proteins that are present in the validation dataset, 10 genes belong to cluster 13, 5 to cluster 12, and 5 to cluster 15 of [10]. Le Roch et al. [10] reported that genes of known functions in clusters 12, 13 are mainly involved in cell-cycle regulation and progression at trophozoite stage, while cluster 15 is characterized as having genes with roles in cell invasion that are under evaluation as blood-stage vaccine. Therefore, hubs of unknown functions in those clusters are more likely to be of these essential functions. It is worth mentioning that, according to the authors of [10], genes from clusters 12 and 13 may represent potential targets for drugs focused on disruption of the highly replicating trophozoite stage of the parasite, while additional candidate vaccine antigens could come from the yet uncharacterized genes of cluster 15. This gives further support to our conjecture that the hubs of unknown functions might be of important biological functions and therefore warrant further investigation.

We believe that the above mentioned evidence demonstrates that the hubs in the constructed malaria gene network tend to be essential. It will also be interesting to investigate those genes among hubs that have not been manually annotated (see Table S3 that contains oligonucleotides of hubs from the overview network with no manual annotation).

4.2.2.   Some Sub-networks in Malaria Gene Network.   It might be interesting to investigate further some sub-networks of the large malaria gene network. As an example, we had a closer look at the glycolytic pathway, as it is mentioned in [3] as the one that is well-preserved in malaria parasite. Among 9 genes from the microarray dataset that belong to this pathway as taken from the http://plasmodb.org database, we found that they share 5 links among themselves. In fact, the probability of 9 randomly picked genes to have 5 links is 0.01% given the connectivity matrix. Given that some of the genes in this pathway are not present in the dataset, this result is encouraging. Our analysis did not pick up MAL61.160 as part of the glycolytic pathway. Instead, another putative copy, PF10_0363, was identified as a part of it, having 2 connections, as well as gene PF10_0155 that has 4 connections.

As another example, we had a look at all major candidates for vaccination (AMA1, EBA175, MSP1, MSP3, MSP7, RAP1, RESA1) studied in [3]. All these genes are very well positioned in the network, having connectivities between 20 and 40, well above the average connectivity of $k = 15$. Interestingly, these vaccine candidates are connected among themselves as well as with some other merozoite invasion proteins (MSP6, MSP8). In addition, the neighbours

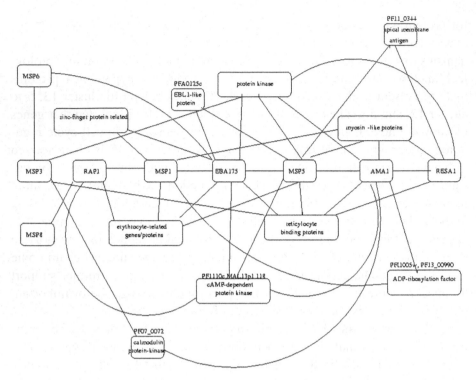

Figure 5. Sketch of a sub-network of seven major malaria vaccine candidates. The sub-network contains major vaccine candidates (AMA1, EBA175, MSP1, MSP3, MSP7, RAP1, RESA1) that have been studied in [3] and some genes/proteins or groups according to our model. Small boxes contain one gene/protein and larger boxes contain two or more related genes/proteins. Only some links are shown. For a full list of links of the seven major malaria vaccine candidates according to our model see Tables S4 on supplemental page.

of these vaccine candidates are enriched with myosin-like proteins, erythro-cyte associated proteins, reticulocyte binding proteins, and zinc finger pro-teins among others (Figure 5). For example, the erythrocyte-related group con-tains erythrocyte binding antigenes (PF07_0128, MAL13P1.60), erythrocyte surface antigene (PFA0110w), and erythrocyte binding proteins (PF08_0142, PF08_0147). The myosin-like proteins group contains 4 genes (PF13_0233, PFL225w, MAL6P1.286, PFL1435c). There are 4 genes in the reticulocyte-binding proteins group (PF13_0198, PFL2520w, MAL13P1.176, PFD0100w). The protein-kinase group includes PF130815w, PFC0945w, PFB00150c, and Ser/Thr protein-kinase PFB0665w. The zinc-finger related group contains one zinc-finger protein (PFE0895) and a cell-cycle regulator with zinc-finger do-main (PFE1415w). There are a large number of hypothetical proteins that are linked to the vaccine candidates in our network. Several of the hypothetical proteins from the list are linked to two major vaccine candidates, while some hypothetical proteins (e.g., PF10_0352, PF07_0127, PFE0365c, PFC1045c, PFD0715c) have links with three major vaccine candidates and are probably worth having a closer look at. For a full list of the neighbours of these major vaccine candidates see Table S4.

5. CONCLUSIONS

In this paper we have constructed a model of malaria gene expression network by a novel method of thresholding two types of pair-wise correlation coefficients: the Pearson correlation and the full-order partial correlation coefficients. The values for thresholds were determined by topological considerations. Both types of correlations are essential in revealing the connections of genes in the network. The constructed small-world, scale-free network has hub-genes that tend to have essential cell functions, similar to other biological networks. We propose that hubs with unknown functions warrant further investigation in the search for malaria vaccine. Finding hubs in the malaria gene network is extremely important in guiding the search for the malaria vaccine. Targeting a highly connecting node with a drug will result in inactivation of a protein that could be fatal to the whole life-cycle of the malaria parasite, whereas removing a less connected node will barely affect the whole system.

This model of malaria gene network is worth investigating further by looking at various sub-networks consisting of genes that are known to be involved in the same biological processes. Alternatively, one might want to look at the neighbours of genes with unknown functions. This might help the process of assigning putative functions to these genes. The links adjacency matrix of the network studied in this paper can be found on the supplemental web-page: www.stats.gla.ac.uk/~raya/Malaria/suppldata.html. To summarise, the thresholding approach of two correlation coefficients that is proposed in this paper suffices for the goal of studying statistical properties of a biological network and also gives encouraging proof-of-principle results.

REFERENCES

[1] Barabasi, A.L. and Oltvai, Z.N., Network biology: Understanding the cell's functional organization, *Nat. Rev. Genet.*, **2** (2004), 101–113.
[2] Bergmann, S., Ihmels, J., and Barkai, N., Similarities and differences in genome-wide expression data of six organisms, *PLoS Biol.*, **2**(1) (2004), E9.
[3] Bozdech, Z., Llinas, M., Pulliam, B.L., Wong, E.D., Zhu, J., and DeRisi, J.L., The transcriptome of the intraerythrocytic developmental cycle of *Plasmodium falciparum*, *PLoS Biol.*, **1**(1) (2003), E5.
[4] Carter, S.L., Brechbuhler, C.M., Griffin, M., and Bond, A.T., Gene expression network topology provides a framework for molecular characterization of cellular state, *Bioinformatics*, **20** (2004), 2242–2250.

[5] de la Fuente, A., Bing, N., Hoeschele, I., and Mendes, P., Discovery of meaningful associations in genomic data using partial correlation coefficients, *Bioinformatics*, **20** (2004), 3565–3574.

[6] Florens, L., Washburn, M.P., Raine, J.D., Anthony, R.M., Grainger, M., Haynes, J.D., Moch, J.K., Muster, N., Sacci, J.B., Tabb, D.L., Witney, A.A., Wolters, D., Wu, Y., Gardner, M.J., Holder, A.A., Sinden, R.E., Yates, J.R., and Carucci, D.J., A proteomic view of the *Plasmodium falciparum* life cycle, *Nature*, **3**(419) (2002), 520–526.

[7] Ideker, T., Thorsson, V., Ranish, J.A., Christmas, R., Buhler, J., Eng, J.K., Bumgarner, R., Goodlett, D.R., Aebersold, R., and Hood, L., Integrated genomic and proteomic analyses of a systematically perturbed metabolic network, *Science*, **929**(5518) (2001), 929–934.

[8] Jeong, H., Tombor, B., Albert, R., Oltvai, Z.N., and Barabasi, A.L., The large-scale organization of metabolic networks, *Nature*, **407**(6804) (2000), 651–654.

[9] Khanin, R. and Wit, E., How scale-free are biological networks, (2005), submitted. www.stats.gla.ac.uk/~raya/howscalefree/howscalefree.pdf

[10] Le Roch, K.G., Zhou, Y., Blair, P.L., Grainger, M., Moch, J.K., Haynes, J.D., De La Vega, P., Holder, A.A., Batalov, S., Carucci, D.J., and Winzeler, E.A., Discovery of gene function by expression profiling of the malaria parasite life cycle, *Science*, **301**(5639) (2003), 1503–1508.

[11] Luscombe, N.M., Babu, M.M., Yu, H., Snyder, M., Teichmann, S.A., and Gerstein, M., Genomic analysis of regulatory network dynamics reveals large topological changes, *Nature*, **431** (2004), 308–312.

[12] Magwene, P.M. and Kim, J., Estimating genomic coexpression networks using first-order conditional independence, *Genome Biology*, **5** (2004), R100.

[13] Przytycka, T. and Yu, Y.-K., Scale-free networks versus evolutionary drift, *Comput. Biol. Chem.*, **28** (2004), 257–264.

[14] Schafer, J. and Strimmer, K., An empirical Bayes approach to inferring large graphical Gaussian models from microarray data, *Bioinformatics*, in press. http://www.stat.uni-muenchen.de/~strimmer/publications/largeggm2004.pdf

[15] Stumpf, M.P., Wiuf, C., and May, R.M., Subnets of scale-free networks are not scale-free: Sampling properties of networks, *Proc. Natl. Acad. Sci. USA*, **102**(12) (2005), 4221–4224.

[16] van Noort, V., Snel, B., and Huynen, M.A., The yeast coexpression network has a small-world, scale-free architecture and can be explained by a simple model, *EMBO Rep.*, **5**(3) (2004), 280–284.

[17] Wille, A., Zimmermann, P., Vranova, E., Furholz, A., Laule, O., Bleuler, S., Hennig, L., Prelic, A., von Rohr, P., Thiele, L., Zitzler, E., Gruissem, W., and Buhlmann, P., Sparse graphical modeling of the iroprenoid gene network in *Arabidopsis thaliana*, *Genome Biology*, **5** (2004), 11.

Chapter 7

Detecting Network Motifs in Gene Co-expression Networks Through Integration of Protein Domain Information

Xinxia Peng[a], Michael A. Langston[b], Arnold M. Saxton[c],
Nicole E. Baldwin[b] and Jay R. Snoddy[a]

[a]*Graduate School of Genome Science and Technology, The University of Tennessee. Oak Ridge National Laboratory, Oak Ridge, TN 37831, USA*
pengxn@ornl.gov, snoddyj@ornl.gov
[b]*Department of Computer Science, The University of Tennessee, Knoxville, TN 37996, USA*
langston@cs.utk.edu, baldwin@cs.utk.edu
[c]*Department of Animal Science, The University of Tennessee, Knoxville, TN 37996, USA*
asaxton@utk.edu

Abstract

Biological networks can be broken down into modules, groups of inter-acting molecules. To uncover these functional modules and study their evolution, our research groups are developing graph-theory based strategies for the analysis of gene expression data. We are looking for groups of completely connected subgraphs (e.g., cliques) in co-expression networks in which corresponding members (genes) encode proteins with the same combination of protein domains. The common pattern shown by a group of such cliques is a "network motif" that may be reused multiple times within organisms. We have developed algorithms for constructing gene co-expression networks labeled with corresponding protein sequence domain combinations, and then detected recurring network motifs with similar protein domain memberships within these labeled networks. The statistical significance of detected network motifs is evaluated by comparing results with those from randomized networks. Also the biological relevance of network motifs is evaluated for shared Gene Ontology annotations on biological processes. We applied our approach to the malaria transcriptome and found many network motifs with three, four, or five members. Many predicted network motifs were further supported by their existence in yeast protein interaction networks. These results illustrate a new strategy for studying the modularity of biological networks by integrating different types of data and cross-species comparisons. A full description of results is available at http://mouse.ornl.gov/~xpv/camda04/.

Keywords: graph algorithms, microarray analysis, clustering, network motif, gene expression, protein domain, protein interaction, data integration

1.   INTRODUCTION

Gene expression microarrays provide a revolutionary approach for measuring the mRNA levels of thousands of genes at the same time. Systematic analysis of genome-wide expression profiles across multiple conditions, together with integration with other kinds of data, should help us gain insight into biological networks. Functionally related genes could be clustered together based on similar expression profiles. Additional information such as Gene Ontology (GO) can typically be exploited to help further biological interpretation of obtained clusters if the target organism is well studied such as yeast, mouse and human, but this data is often not sufficiently available for other important organisms. General clustering algorithms, moreover, produce clusters of relatively large size, making it difficult to test the clusters of interest using wet-lab experiments. In addition, general clustering algorithms do not provide reasonably detailed information about the relationship among genes in a cluster, such as if some genes directly interact with each other and how. This makes it even more difficult for individual researchers to verify the associations among genes predicted by clustering algorithms experimentally.

Additional independent information is needed to break big clusters into smaller ones and thereby provide more detailed insights into relationships among genes within subclusters. Protein sequence information is a good candidate. Proteins can be decomposed into protein domains, both the units of protein function and evolution. More importantly, there is considerable evidence that biological systems build various functional units by reusing protein domains in different combinations [10]. We can hence attempt to decode the common mechanisms used in biological systems through studying protein domains.

Duplication and divergence are important components in the evolution of genomes and biological complexity. Duplicated genes can retain or change their interaction partners. They may, over time, replace interaction partners, but the duplicated gene might still interact directly or indirectly with a partner having similar characteristics to the original partner. Multiple instances of MAP3K-MAP2K-MAPK three-tiered cascades constitute a well studied example [4]. It is still unknown whether it is a general principle in biology that different genes form instances of common patterns such as in MAPK pathways.

In this study, we developed novel algorithms to decompose the clusters of genes into smaller ones by integrating protein domain information into the clustering algorithm. Our algorithm is able to provide more detailed information about putative relationships among genes within clusters by examining corresponding protein domain functions. In addition, we provide evidence that some units of similar function are temporally regulated differently at the transcriptional level. To increase confidence, our approach is able to integrate additional information, such as protein interaction data from different species. Yeast is a good source because rich information has been already collected.

2. MATERIALS AND METHODS

2.1. Co-expression Networks

In a co-expression network, the genes are represented by vertices (nodes). An unweighted and undirected edge (connection) is placed between two genes if they are co-expressed, as determined by having a correlation higher than some specified threshold. A malaria transcriptome expression data set [3] was downloaded from the CAMDA04 website (http://www.camda.duke.edu/camda04/datasets/) and the Complete Dataset was used in this study. All Cy5/Cy3 ratio intensities were log 2 transformed. For ORFs represented by multiple oligonucleotides on the DNA microarray, the expression ratios were averaged. Gene pairwise correlation coefficients were calculated using the standard Pearson method. Correlation coefficients between pairs of genes computed with fewer than 33 of 46 timepoints (approximately 75%) due to missing values were discarded. The final correlation matrix had 3,842 unique ORFs after removing those genes which did not share 33 or more non-missing datapoints with at least one other gene. Based on a selected cutoff value, the calculated correlation matrix was converted into a binary symmetric matrix of the same size. An entry in this matrix was set to 1 if the corresponding correlation coefficient was greater than or equal to the cutoff value, otherwise the entry was set to 0. Rows (columns) with all-zero entries were deleted from the binary matrix, corresponding to the elimination of isolated vertices in the graph associated with such a matrix.

2.2. Protein Domain Annotation

Plasmodium falciparum protein sequences and GO annotations were downloaded from PlasmoDB (http://plasmodb.org). To get protein domain annotations, all protein sequences were searched against Pfam HMM library (Release 14.0, global, ls mode, Pfam-A HMMs with a total of 7459 families) using hmmpfam, a program provided by HMMER package. The trusted cutoffs built in Pfam library were employed. The HMM library was downloaded from Pfam website (http://www.sanger.ac.uk/Software/Pfam). HMMER 2.3.2 was downloaded from http://hmmer.wustl.edu. The computation was done on the OIT Cluster of 32 nodes of the SInRG project (http://icl.cs.utk.edu/sinrg/index.html).

2.3.   Network Motif Discovery

The concept of "network motifs" was first proposed by Alon's group in studying various real world networks including biological networks [8,11]. Network motifs were defined as patterns of interactions recurring more frequently in a network than in randomized networks. Here we extended the concept of network motif to labeled graphs by studying patterns of vertex labels (Figure 1). As shown in Figure 1A, a hypothetical network motif might be a clique of three genes. These genes are highly co-expressed as required by the correlation cutoff to create an edge. In addition, each gene has its own characteristic protein domain information as reflected in its label. Figure 1B shows 7 hypothetical genes in a co-expression network forming three distinct instances of the network motif as described in Figure 1A. In each of the instances, three genes are highly correlated with each other as indicated by the edges, and their protein domain information maps one-to-one to the specified network motif based on rules as described below. Among the three instances, instances II and III share at least one gene (here two genes) and we say these two instances are "overlapping". On the other hand, instance I does not share any genes with instance II, so these two are "non-overlapping". Instances I and III form another pair of non-overlapping instances. In general, we insist that in a network motif

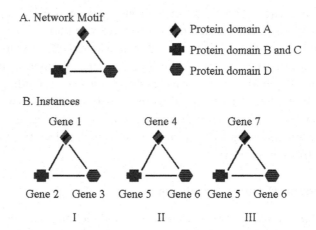

Figure 1. Schematic of network motif. (A) A network motif is a pattern as a complete connected subgraph (e.g., cliques) of certain size k ($k = 3$ here) and the vertices are labeled as reflected by the shapes and shadings. Here vertices represent genes and the vertex labels are the protein domain information of the proteins encoded by the corresponding genes. (B) Seven hypothetical genes in a co-expression network forming three distinct instances of the network motif as described in (A). In each of the instances, three genes are highly correlated with each other as indicated by the edges, and their protein domain information maps one-to-one to the network motif. Instances II and III share at least one gene (here two genes) and we say these two instances are "overlapping". Instance I does not share any genes with instance II, so these two are "non-overlapping". Instances I and III form another pair of non-overlapping instances.

of size k ($k = 3$ in the above example) every pair of vertices is joined by an edge, that is, the network motif forms a clique.

Starting from the above calculated *P. falciparum* co-expression network, we converted it into a labeled graph whose vertices (genes) were labeled with their corresponding annotated Pfam protein domain information. The clique-based clustering algorithm of [6] was applied to this labeled co-expression network to search for patterns of highly co-expressed genes or network motifs (Figure 1). For a specified k, we scanned all k-vertex cliques and grouped all cliques found based on the protein domain information. Within a group of cliques, protein domain information in each clique can match one-to-one against protein domain information of genes in any other clique. These groups of cliques are called "putative network motifs". Next, a parameter f specifying the minimum number of mutual non-overlapping instances in a network motif was used to trim the list of putative network motifs. Only putative network motifs having at least f non-overlapping instances were kept as network motifs.

To account for the abundances of different domains in the whole genome, we further assessed the statistical significance of each detected network motif by comparison to randomized networks. Starting from the real co-expression network, we generated a randomized network by randomly permuting the domain labels of all genes while leaving the connection structure of the graph untouched, and then ran the same network motif detection procedure on the resulting randomized network. This process was repeated 1,000 times. The fraction of times the same network motif was found in the randomized networks was defined as the p-value for the network motif.

Matching of protein domain information between two genes can be classified into many possible levels, but here we propose only domain matching levels A and B. Level A requires that two proteins have the exact same type of domain, the same number of each type of domain and all domains in the same order in the respective protein sequences from N-terminal to C-terminal. Domain matching level A is a global alignment that suggests that the two proteins are essentially the same in terms of domain architecture. Domain matching level B only requires the same types of domain, with no constraints on the number and the order of domains in the proteins. At this level, the domain duplication and domain shuffling during evolution are permitted while suggesting that the basic molecular functions of each protein might be similar. The network motif detection procedure was run separately using different domain matching levels.

2.4.   Protein Interaction Networks

A yeast protein interaction dataset was downloaded from the BIND website (http://www.blueprint.org/bind/bind.php). In this protein interaction network, genes were again represented by vertices (nodes). An un-weighted and undirected edge (connection) is placed between two genes if there is a documented interaction between these two genes. Since the topologies of most protein complexes are unknown at this time, we converted protein complexes into binary interactions using the "matrix" model, which put edges between all possible pairs of genes in the same protein complex [1]. The use of the matrix model facilitates searching for possible instances of network motifs found in co-expression networks in protein complexes. Yeast GO annotations were downloaded from SGD (http://www.yeastgenome.org/).

2.5.   Data Visualization

Detected network motifs are presented on the web using ALIVE (http://mouse.ornl.gov/alive). Expression plots were drawn using R (http://www.r-project.org).

3.   RESULTS

3.1.   Co-expression Networks

To convert a correlation matrix into a corresponding binary co-expression network, a suitable cutoff value for the correlation coefficient must be chosen. Based on the previous reports that biological networks, including co-expression networks, follow a scale-free distribution of connectivities [2,7], we chose a cutoff value which gave fewer vertices with higher degree (connectivity). Plots of the degree distribution for graphs generated under a series of cutoff values suggest that a correlation cutoff value of 0.95 is appropriate through visual inspection (Figure 2).

This value was surprisingly higher than our expectation. We compared the distribution of correlation coefficients of this dataset with those of several cell/life cycle gene expression datasets and found the distribution of correlations in this dataset showed a characteristic bimodal shape while others had bell-like shape (data not shown). One of the possible reasons is that the majority of genes in this dataset exhibit periodicity [3]. Within this data set and others, we observed that genes which exhibit periodicity tend to shift the distribution toward higher correlations. When the genes in the Overview Dataset that were selected based on their strong periodic behavior were removed, the degree distributions of those resulting networks did tend to have fewer vertices of higher degrees compared to the original networks (see online supplement). We further verified that the resulting co-expression network ($R \geqslant 0.95$) were enriched (p-value < 0.001, chi-square test) with genes of periodic behavior. About 93% (2,124 of 2,292) of unique ORFs in the co-expression network ($R \geqslant 0.95$) are in the Overview Dataset of 2,714 ORFs (about 78%) while

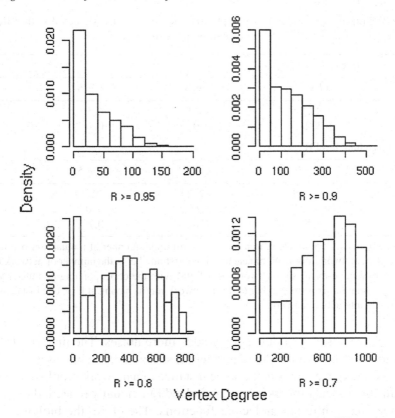

Figure 2. Degree distribution of co-expression networks generated under different cutoff values of correlation coefficient (R). For each cutoff value (shown at the bottom of each plot), a co-expression network was generated (see the main text for details), and the histogram of the degrees of all vertices (the numbers of connections of vertices) was plotted using R with default settings. The horizontal axis is the vertex degrees and the vertical axis is the relative frequencies.

only about 36% (559 out of 1550) of those genes removed by this cutoff were in the Overview Dataset.

3.2. Prediction of Network Motifs

Using a series of values for parameters k, the size of network motifs and f, the minimum number of non-overlapping instances, we found a number of putative network motifs under different domain matching levels (Table 1). As shown in Table 1, both increasing k and f decrease the number of network motifs detected (first number in each cell). More network motifs were found at domain matching level B than at level A with the same corresponding parameter values for k and f, probably because of the less stringent constraints on matching protein domain information. More studies are needed to check if a better coverage is achieved at domain matching level B by including more dis-

Table 1. Summary of the number of putative network motifs detected with different set of parameter values

Domain matching level A

	$k = 3$	$k = 4$	$k = 5$	$k = 6$
$f = 2$	88, 25	18, 11	6, 5	1, 1
$f = 3$	3, 2	0, 0	0, 0	0, 0

Domain matching level B

	$k = 3$	$k = 4$	$k = 5$	$k = 6$
$f = 2$	197, 53	87, 29	32, 17	9, 6
$f = 3$	17, 13	6, 6	0, 0	0, 0
$f = 4$	5, 5	0, 0	0, 0	0, 0

k: the size of network motifs to search for. f: the minimum number of mutual non-overlapping instances for a network motif. Within each cell the first number is the number of network motifs found in malaria co-expression network ($R \geqslant 0.95$) with the corresponding parameter values, and the second number is the number of those network motifs having at least one instance in the yeast protein interaction network.

tantly related genes, or it just simply adds more noises. The majority (>95%) of putative network motifs have p-values less than 0.05 in all cases.

We assumed that genes in the same instance of a network motif should share the same biological process if they are indeed functionally related, though each gene may have different molecular functions. Therefore, the biological relevance of the putative network motifs was evaluated by simply counting the number of genes within an instance that share the same GO terms in a biological process category. Although the GO annotations on malaria genes are relatively limited, we can still observe that genes with GO annotations in the same motif instance did tend to share similar terms. We also used the functional gene groups as provided in [3] to check the similarity of functions of genes in the same instances, and this gave similar results.

Figure 3A shows a putative network motif detected under domain matching level A, $k = 6$ and $f = 2$. This motif consists of six highly co-expressed genes. Three of six genes have the same domain combination as two domains ordered from N-terminal to C-terminal, DEAD/DEAH box helicase (PF00270) and Helicase conserved C-terminal domain (PF00271). These genes are involved in various aspects of RNA metabolism as suggested by the Pfam domain annotation. One of the six genes has three WD domains, G-beta repeats (PF00400), one has a Brix domain (PF04427) and the last one has GTPase of unknown function (PF01926). The protein domain functions suggest that this network motif is involved in ribosome biogenesis [5,9]. Figure 3B shows the *P. falciparum* genes form various instances of the network motif through different combinations of genes. (Genes are shaded in

A. Network motif   B. Instances

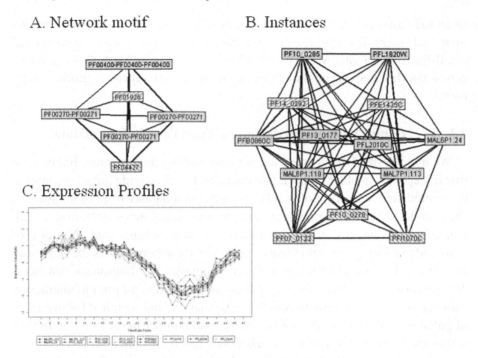

C. Expression Profiles

Figure 3. An example network motif. (A) A putative network motif detected at domain match-ing level A with parameter values $k = 6$ (size) and $f = 2$ (minimum number of mutual non-overlapping instances). This network motif consists of six highly co-expressed genes and three of them have the same domain combination as two domains ordered from N-terminal to C-terminal, DEAD/DEAH box helicase (PF00270) and Helicase conserved C-terminal do-main (PF00271). One of the six genes has three WD domains, G-beta repeats (PF00400), one has a Brix domain (PF04427) and the last one has GTPase of unknown function (PF01926). (B) Thirteen *P. falciparum* genes form various instances of the network motif through different combinations of genes. Genes are shaded in the same way as those Pfam accession numbers shown in (A) to indicate their corresponding domain information. (C) The expression profiles of those 13 genes.

the same way as those Pfam accession numbers shown in Figure 3A to in-dicate their corresponding domain information.) Only 5 of 13 genes were assigned with functional group annotation and all of these five genes were in Cytoplasmic Translation Machinery functional group [3]. Only 3 of the 13 genes have GO annotations. Significantly, these three genes are a sub-set of the group of five and they all were assigned with the same GO terms as RNA metabolism (GO:16070), nucleobase, nucleoside, nucleotide and nu-cleic acid metabolism (GO:6139), cell growth and/or maintenance (GO:8151) and metabolism (GO:8152). These GO annotations are very broad, but agree with the more specific hypothesis that these genes are related to ribosome bio-genesis. This group of 13 genes may potentially work together in some way, since they all had very similar expression profiles under these diverse developmental stages. It is also possible that these genes were organized as several small functional units. More information is needed to dissect this cluster of genes, but this analysis might suggest some initial inferences to guide experiments.

3.3.   Confirmation of Prediction by Yeast Protein Interactions

We hypothesized that a predicted network motif would be more likely to be true if it appears in other independent datasets. It will provide further support to those predicted network motifs if they appear in a dataset from other species, since protein–protein associations may be transferred across organisms [12]. One of the advantages of treating protein domains as functional units of proteins and labeling genes with their protein domain information is the flexibility of doing cross-species comparisons across significant evolutionary distances. To gain further confidence in our predictions, we used yeast protein interaction data that includes rich protein complex information, and searched for instances of putative malaria network motifs. The second number in each cell of Table 1 shows the number of malaria network motifs having instances in yeast protein interaction network. We can see that relatively more malaria network motifs were supported by yeast interaction data as the parameters became more stringent. The matrix model increased the coverage of the yeast proteome by including all possible true interactions within the experimental data, but some false interactions were also added [1]. We felt this model was the best compromise for module discovery. The results from the protein interaction data should be further studied individually, especially when there is better experimental verification.

Figure 4 shows the instances formed by different combinations of 27 genes detected in the yeast protein interaction network for the malaria network motif shown in Figure 3. Forty-five protein complexes stored in the BIND database have at least two members belonging to this group of 27 genes. This strongly suggests that these gene products directly interact with each other under different conditions in various ways. One extreme example is that protein complex 11635 contains six genes forming an exact instance of the predicted network motif. The two largest groups of genes sharing a common GO annotation in this group of 27 genes are a group of 9 genes annotated as ribosomal large subunit assembly and maintenance (GO:27) and the other 8 genes as 35S primary transcript processing (GO:6365). These two groups totally cover 13 out of 27 genes. All of the evidence above suggests that this particular network motif represents a core interaction unit for various protein complexes involving cytoplasmic translation, or even more specifically as ribosome biogenesis.

A. Network motif   B. Instances

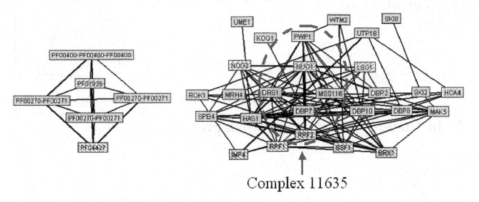

Complex 11635

Figure 4. Instances found in the yeast protein interaction network for the network motif shown in Figure 3. (A) The similar yeast network motif as shown in Figure 3. (B) A subgraph of the yeast protein interaction network was shown, and only included vertices (genes) forming at least one instance of the network motif. Yeast genes are shaded in the same way as in Figure 3A to indicate their corresponding domain information. Highlighted is an instance of the network motif in which all six proteins present in the protein complex 11635 in BIND database.

This hence supports tentative functional assignment of all the involved malaria genes that had no prior annotation. The strength of our strategy is both to cluster functionally related genes and to provide more detailed information about relationships among these genes by integrating information from multiple orthogonal sources.

3.4.   Prediction of Complementary Functional Units

A network motif represents a specific combination of individual protein domains, and this combination can carry out a special function shared by individual instances as relatively independent subsystems. We hypothesized that individual instances of a network motif could function in different locations and times, dependent upon regulation. The malaria time series data enables us to test this hypothesis by examining the temporal expression profiles of instances of network motifs. Figure 5 shows such an example network motif detected at domain matching level B with parameter values $k = 3$ and $f = 2$. This network motif represents a combination of three independent domains, AhpC/TSA family (PF00578), protein kinase domain (PF00069) and Calcineurin-like phosphoesterase (PF00149) (Figure 5A). Six *P. falciparum* genes form two independent instances of this network motif (Figure 5B). The AhpC/TSA family contains Peroxiredoxins (Prxs), a ubiquitous family of antioxidant enzymes and Prxs can be regulated by phosphorylation [13]. The paired kinase and

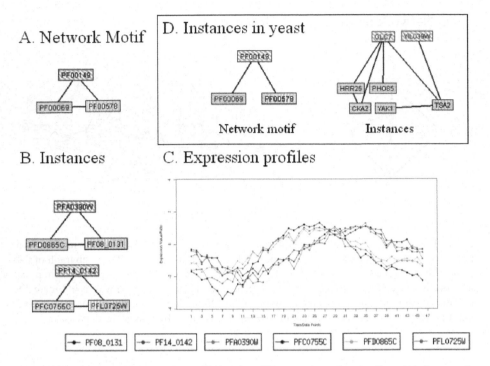

Figure 5. Instances of a network motif showing different expression profiles. (A) A network motif detected at domain matching level B with parameter values $k = 3$ and $f = 2$. This network motif represents a combination of three independent domains, AhpC/TSA family (PF00578), protein kinase domain (PF00069) and Calcineurin-like phosphoesterase (PF00149). (B) Six *P. falciparum* genes form two instances of this network motif. (C) Expression profiles of these six genes. (D) Instances of the network motif were found in yeast protein interaction network.

phosphatase may reflect that these two Prxs are tightly controlled through phosphorylation and dephosphorylation. Of striking interest is that apparently these six genes all have similar expression profiles and the only major difference is the timing. There is a phase difference between two instances while all three genes within each of two instances have very similar timing. When these expression profiles are compared with morphological data [3], we would conclude that one instance (PF08_0131, PFD0865c, PFA0390w) functions at trophozoite stage and another (PF14_0142, PFC0775c, PFL0725w) at schizont stage based on their peak expression values. Having instances in yeast protein interaction data provides further support that these genes do interact directly (Figure 5D). It is worth mentioning that none of these genes were assigned to a functional group [3] and these six genes share very broad GO annotations such as cell growth and/or maintenance (GO:8151) and metabolism (GO:8152).

4. DISCUSSION

With the rapid development of high-throughput methods such as microarrays in recent years, massive amounts of experimental data have been collected for different species under various conditions. New computational approaches are needed to analyze these data in an integrative way and provide more reliable results with finer resolution for experimental verification. Here we propose a new strategy to analyze gene expression data by integrating a diversity of additional information, such as primary sequence information and protein interaction data. Though the present study starts with a dataset from a single species, our approach is used here to look for patterns in other species and can be easily generalized to begin with information from multiple species.

The strategy of integrating protein domain information into expression data analysis was based on our hypothesis that genes/proteins form relatively independent functional modules. Gene expressions in these modules will be well coordinated because of selective forces or functional constraints. The possible origins of these modules are gene duplication and reuse of protein domains. This then implies that these modules might form some common patterns at protein domain level that we can observe in experimental data. Our relatively detailed predictions of association among genes as shown in Figure 5 provide rich information for experimental verification and elucidation by examining other data in light of these presumptive network motifs, including studies of networks under other conditions. Several things can be done to refine the analyses. For example, we might be able to combine instances of these motifs together toward building up larger network components such as large pathways or protein complexes. We should be able to loosen the strict requirements for exact cliques for motifs of interest, as evolution will not always preserve exact co-expression matches and not all members of a motif will duplicate. The general approach that we have begun to outline here, we believe, can become a useful tool to ask a number of other interesting research questions about how networks work in the present time and how they arose to work that way over evolutionary time.

An older version of this paper, but with color figures, is available as a technical report as University of Tennessee Computer Science Technical Report UT-CS-05-545. See: http://www.cs.utk.edu/~library/tech_reports.php.

ACKNOWLEDGEMENTS

We thank three anonymous reviewers for their careful reading of the manuscript and helpful suggestions that have improved the manuscript. Research sponsored, in part, by the Office of Biological and Environmental Research, managed by UT-Battelle, LLC for the U.S. Department of Energy under Contract No. DE-AC05-00OR33735 and, in part, by National Institute of Health grants to the University of Tennessee, including NIDA project No. 5-P01-DA015027-02 and NIAA project No. 5-U01-AA012512-02. This research has used resources provided by the Scalable Intracampus Research Grid (SInRG), a project at the University of Tennessee supported by the National Science Foundation under award EIA-9972889. It has also been supported by the Office of Naval Research under grant N00014-01-1-0608, by the Department of Energy under contract DE-AC05-000R22725, and by the Tennessee Center for Information Technology Research under award E01-0178-261.

REFERENCES

[1] Bader, G.D. and Hogue, C.W., Analyzing yeast protein–protein interaction data obtained from different sources, *Nat. Biotechnol.*, **20**(10), 991–997.

[2] Bhan, A., Galas, D.J., and Dewey, T.G., A duplication growth model of gene expression networks, *Bioinformatics*, **18**(11), 1486–1493.

[3] Bozdech, Z., Llinas, M., Pulliam, B.L., Wong, E.D., Zhu, J., and DeRisi, J.L., The transcriptome of the intraerythrocytic developmental cycle of *Plasmodium falciparum*, *PLoS Biol.*, **1**(1), E5.

[4] Chang, L. and Karin, M., Mammalian MAP kinase signalling cascades, *Nature*, **410**(6824), 37–40.

[5] Eisenhaber, F., Wechselberger, C., and Kreil, G., The Brix domain protein family – a key to the ribosomal biogenesis pathway? *Trends in Biochemical Sciences*, **26**(6), 345–347.

[6] Langston, M., Lin, L., Peng, X., Baldwin, N., Symons, C., Zhang, B., and Snoddy, J., A combinatorial approach to the analysis of differential gene expression data: The use of graph algorithms for disease prediction and screening, in: *Methods of Microarray Data Analysis IV*, Springer-Verlag, New York.

[7] Lee, H.K., Hsu, A.K., Sajdak, J., Qin, J., and Pavlidis, P., Coexpression analysis of human genes across many microarray data sets, *Genome Research*, **14**(6), 1085–1094.

[8] Milo, R., Shen-Orr, S., Itzkovitz, S., Kashtan, N., Chklovskii, D., and Alon, U., Network motifs: Simple building blocks of complex networks, *Science*, **298**(5594), 824–827.

[9] Neer, E.J., Schmidt, C.J., Nambudripad, R., and Smith, T.F., The ancient regulatory-protein family of WD-repeat proteins, *Nature*, **371**(6495), 297–300.

[10] Pawson, T. and Nash, P., Assembly of cell regulatory systems through protein interaction domains, *Science*, **300**(5618), 445–452.

[11] Shen-Orr, S.S., Milo, R., Mangan, S., and Alon, U., Network motifs in the transcriptional regulation network of *Escherichia coli*, *Nature Genetics*, **31**(1), 64–68.

[12] von Mering, C., Jensen, L.J., Snel, B., Hooper, S.D., Krupp, M., Foglierini, M., Jouffre, N., Huynen, M.A., and Bork, P., STRING: Known and predicted protein–protein associations, integrated and transferred across organisms, *Nucleic Acids Res.*, **33**, Database Issue, D433-7.

[13] Wood, Z.A., Schroder, E., Robin Harris, J., and Poole, L.B., Structure, mechanism and regulation of peroxiredoxins, *Trends Biochem. Sci.*, **28**(1), 32–40.

Chapter 8

Chromosomal Clustering of Periodically Expressed Genes in *Plasmodium falciparum*

Pingzhao Hu[a], Celia M.T. Greenwood[a,b], Cyr Emile M'lan[c] and Joseph Beyene[a,b]

[a]*Hospital for Sick Children Research Institute, University of Toronto, 555 University Avenue, Toronto, ON, M5G 1X8 Canada*
[b]*Department of Public Health Sciences, University of Toronto, 555 University Avenue, Toronto, ON, M5G 1X8 Canada*
[c]*Department of Statistics, University of Connecticut, Storrs, CT, 555 University Avenue, Toronto, ON, M5G 1X8 Canada*

Abstract

Identification of periodically expressed genes (PEGs) has been widely studied, but understanding how PEGs are distributed along chromosomes is largely unexplored. In this study we investigated chromosomal clusters of PEGs in stages of intraerythrocytic developmental cycle (IDC) of *Plasmodium falciparum* using the cDNA microarray data provided by the organizers of the Critical Assessment of Microarray Data Analysis (CAMDA) 2004 competition. To this end, we implemented an analysis consisting of three stages: first, fitting sinusoidal curves to the 46 time points to identify periodically expressed oligonucleoitides, second, using a support vector machine (SVM) to assign the periodically expressed oligonucleoitides to the four known developmental stages of the IDC, and third, defining stage-specific physically adjacent clusters and evaluating through permutation whether there were more clusters than expected by chance. We identified 2949 periodically expressed oligonucleoitides (2204 genes) where periodicity explained at least 70% of the variation over time, and 718, 624, 141, and 167 genes were assigned to the ring/early trophozoite, trophozoite/early schizont, schizont, and early ring stages, respectively, with at least 80% probability for stage prediction. Finally, we identified 312 clusters of two or more adjacent genes assigned to the same stage. Using a permutation-based method, we found that we observed more clusters of size five than expected by chance ($p = 0.04$). There was also a suggestion ($p \sim 0.10$) of more clusters than expected for other cluster sizes. Our findings suggest that the expression of periodically expressed genes may be coordinated locally on chromosomes where there are clusters of genes within same stage, suggesting cis-regulation.

Keywords: asexual intraerythrocytic development cycle, multiple linear regression model, support vector machine, class probability, chromosomal clusters

1.   INTRODUCTION

Plasmodium falciparum is one of the organisms that cause human malaria. The 22.8 Mb genome of *P. falciparum* is comprised of 14 linear chromosomes. Understanding the genome of *P. falciparum* will hopefully provide a foundation for prevention and treatment of the disease. The complete *P. falciparum* life cycle includes three major developmental stages: the mosquito, liver and blood stages. In order to try and identify which genes are active in influencing the development cycle (called the asexual intraerythrocytic development cycle (IDC)), it has been postulated genes that exhibit periodic patterns of expression are likely to be involved in regulating the IDC. Several papers have presented methods for identifying periodically expressed genes (PEGs). For example, Bozdech et al. [4] and others [3,16,19] quantified the periodicity of the expression profile of each gene by using Fourier analysis. Booth et al. [3] implemented a one-component Fourier analysis by using a linear model containing sine and cosine waves. The procedure emphasized the use of standard statistical methods, such as multiple linear regression, together with the R^2 measure of goodness of fit, and F-tests for significance. Lu et al. [12] proposed a more general mixture model where a periodic function (modeled with sine and cosine waves) is convoluted with a normal distribution that allows different cells to be in different phases of the cell cycle. Furthermore, they used a 2-component beta-mixture to model the residual sums of squares and to obtain the probability that a gene is periodically expressed. The straightforward linear model with sine and cosine waves used by Booth et al. [3] provides an easy way to model a periodic function, and the R^2 summary succinctly summarizes how much of the periodicity is explained by the model.

There have been several proposals for assigning PEGs to different cell-cycle stages. Two studies have used unsupervised clustering methods to classify genes into cell cycle phases [16,19]. However, these methods require an arbitrary specification of the number of clusters in a dataset, and furthermore cannot use prior information. Since, for *Plasmodium falciparum*, the stage of action is known for several hundred genes, this information could be better used by a good supervised classification method. Lu et al. [12] calculated the Pearson correlations between gene profiles and "typical" transcription profiles of genes of known stages in order to assign PEGs into stages. Booth et al. [3] grouped the genes by comparing the ratios of the coefficient for the sine wave divided by the coefficient for the cosine wave to genes whose phase was known.

Bozdech et al.'s study [4] showed that the PEGs in *Plasmodium falciparum* are likely to be co-regulated. Previous studies on *Saccharomyces cerevisiae* [7],

Homo sapiens [5] and *Caenorhabditis elegans* [15] have demonstrated that co-regulated genes often cluster together on chromosomes. Proteomic analysis of the three developmental stages of *P. falciparum* also revealed the presence of chromosomal clusters encoding co-expressed proteins [9]. However, the number and composition of clusters will vary substantially with the cluster definition and the experimental design (e.g., marker density) of each study. Ad hoc criteria have been mainly used for identifying chromosomal clusters of coregulated genes, either based on a sliding window of a given distance [4,15] or on a given number of adjacent genes [9]. Interpreting the significance of an observed cluster is therefore challenging.

In this study, we first applied linear model used in Booth et al. [3] to identify periodically expressed oligonucleoitides. Secondly, we assigned oligonucleoitides into IDC stages by using support vector machines (SVMs), a widely used supervised classification algorithm, combined with a statistically rigorous approach for converting the SVM results into probabilities of stage membership [14]. Thirdly, we define stage-specific clusters of genes and use a stage-specific permutation approach to examine whether more clusters, following our definition, are observed than expected by chance. The second and third steps of our analyses provide new insights and analytic strategies that have not been explored in other previous studies.

2.   MATERIALS AND METHODS

2.1.   Data Source and Preprocessing

The organizers of CAMDA 2004 provided three datasets: the complete raw data set, a quality controlled data set and an overview data set. In this study we used the quality-controlled data set to simplify the preprocessing and to facilitate comparisons with the original work on this dataset [4]. The data set includes 5080 oligonucleotides measured at 46 time points spanning 48 hours. The data was originally normalized using the NOMAD (Normalization of MicroArray Data) database system. 243 of the oligonucleotides had a missing value at one or more time points. We imputed missing values in the dataset using the 10-nearest neighbor averaging method [18]. This imputation method can be summarized as follows: if oligonucleotide x has one missing value at time point j, the approach first finds 10 other oligonucleotides that have a value measured at time point j, with expression most similar to x at all other 45 time points using a Euclidean metric. Then the weighted average of expression values for time point j from these 10 similar oligonucleotides is used as an estimate of the missing intensity value in oligonucleotide x. The inverse of the Euclidean distance was used to weight the average.

2.2.   Identification of Periodically Expressed Oligonucleotides

Since many genes were measured by more than one oligonucleotide, we fitted a linear model for the expression profile of each oligonucleotide. For oligonucleotide i at time point j, the variation in log expression ratios over the course of the study was modeled as a linear combination of sine–cosine waves as follows:

$$y_{ij} = b_{0i} + b_{1i} \cos(2\pi t_j / T) + b_{2i} \sin(2\pi t_j / T) + e_{ij}, \tag{1}$$

where T is the period for the cyclically expressed oligonucleotides. We estimated the period by minimizing the sum of squared errors (SSE) of least squares fits of known periodically expressed oligonucleotide profiles to model (1), over different values of T.

Equation (1) is a standard multiple linear regression model, so the regression parameters b_{0i}, b_{1i}, b_{2i} can be estimated using the least squares method, for fixed T. In order to evaluate whether an oligonucleotide is periodically expressed in the intraerythrocytic development cycle, the goodness-of-fit of the linear model for each oligonucleotide's expression profile was measured by R^2. The R^2 value quantifies the "proportion of variance explained (PVE)" by the periodicity. The PVE falls between zero and one, and values close to one indicate greater periodicity for a given T. The statistical significance of each R^2 can be determined by the F-statistic [3], $F = (J - p)R^2 / ((p - 1)(1 - R^2))$. Here J is the number of time points and $p = 3$ is the number of parameters in the linear model.

Selecting periodically expressed oligonucleotides based on F-statistics involves multiple testing as described by Dudoit et al. [8]. The false discovery rate (FDR) [2] has become a popular error measure for controlling the false positive and false negative errors in this situation. We applied Taylor et al.'s algorithm [17], a column-wise permutation-based method (that is, we permuted the times in the data) to calculate the FDR. In their method, T-statistics were used since they were testing for differences between two experimental conditions. To apply a conceptually similar approach to our F-statistics is straightforward, and proceeds as follows:

1. Create B column-wise permutations of the times, fit the linear model in Equation (1), and obtain F-statistics $F_{1,b}, \ldots, F_{I,b}$ testing for periodicity, for oligonucleotide $i = 1, 2, \ldots, I$ and permutations $b = 1, 2, \ldots, B$.
2. Let $F_{i,0}$ be the F-statistics for oligonucleotide i in the original data, let F_c be a chosen cutoff, let $\widehat{R} = \sum_{i=1}^{I} I(|F_{i,0}| \geqslant F_c)$, and let $\widehat{V} = (1/B) \sum_{b=1}^{B} \sum_{i=1}^{I} I(|F_{i,b}| \geqslant F_c)$.

3. Estimate the FDR by $\pi_0 \widehat{V} / \widehat{R}$, where π_0 is the true proportion of oligo-nucleotides without periodicity among all the oligonucleotides I.

We followed Taylor et al.'s methods [17] to calculate π_0. Statistically significant oligonucleotides were chosen by comparing the F-statistic $F_{i,0}$ with a given cutpoint F_c at the estimated FDR.

2.3.   Classification of Periodically Expressed Oligonucleotide

In this study, the IDC is known to contain 4 stages, namely, ring/early trophozoite, trophozoite/early schizont, schizont and early ring, and a total of 472 oligonucleotides (351 genes) are known to be expressed in one of these stages [4]. Based on Table S2 and Figure 2 of the Bozdech study [4], there are 214, 93, 131 and 34 periodically expressed oligonucleotide in these four stages respectively. Therefore, to classify the oligonucleotides identified in Section 2.2 into the four stages with high confidence, we used a pairwise coupling method to solve this multi-class classification problem [11]. This involves estimating class probabilities for each pair of classes, and then coupling the estimates together for each oligonucleotide.

We employed support vector machines (SVM) with a radial basis function (RBF) kernel as our base classifier for each pair of classes. SVM is a core machine learning technique with a strong theoretical basis and excellent empirical success [20]. Generally speaking, given a periodically expressed oligonucleotide x, the SVM outputs a decision value f_{kl} for each pair of classes k and l. While the sign and magnitude of f_{kl} can be used to determine the class prediction and the confidence level of that prediction, the SVM decision value f_{kl} is an uncalibrated value that does not always translate directly to a probability value useful for estimating confidence. Platt [14] proposed a parametric model for calibration in which the class probability r_{kl} for each pair of classes k and l was estimated based on: $\hat{r}_{kl} = \frac{1}{1+e^{A f_{kl}+B}}$, where A and B are estimated by minimizing the negative log-likelihood function.

A common way to combine pairwise comparison scores r_{kl} is through a majority voting method described by Friedman [10]. The voting method selects the class label with the most winning two-class decisions. In our study, however, we required a confidence level in order to assign a periodically expressed oligonucleotide into a stage. Hastie and Tibshirani [11] proposed an algorithm to calculate coupled class probabilities for this task. For the periodically expressed oligonucleotide x, the pairwise calibrated SVM computes estimates $\hat{r}_{kl}$ for classes $k, l = 1, \ldots, 4, k \neq l$. Assume that n_{kl} is the number of genes in the training set for the classifier trained on classes k and l. We wish to estimate $\{p_k\}_{k=1}^4$, where $p_k = p(class = k | x)$. The algorithm of Hastie and Tibshirani works as follows:

1. Start with some initial $\hat{p}_k > 0$, and corresponding $\hat{u}_{kl} = \hat{p}_k/(\hat{p}_k + \hat{p}_l)$.
2. Repeat ($k = 1, \ldots, 4, 1, \ldots$) until convergence:

$$\hat{p}_k \leftarrow \hat{p}_k \frac{\sum_{k \neq l} n_{kl} \hat{r}_{kl}}{\sum_{k \neq l} n_{kl} \hat{u}_{kl}}, \qquad \hat{r}_{kl} = \frac{1}{1 + e^{Af_{kl}+B}},$$

$$\hat{p} \leftarrow \hat{p} \Big/ \sum_{k=1}^{4} \hat{p}_k, \quad \hat{p} = (\hat{p}_1, \hat{p}_2, \hat{p}_3, \hat{p}_4)$$

   recompute the $\hat{u}_{kl}$.
3. The final class prediction y is based on the maximum, $\hat{p}_y \leftarrow$ arg max$_k(\hat{p}_k)$, and so we assign $\hat{p}_y$ as the probability that the oligonucleotide x falls into the predicted stage $y \in \{1, 2, 3, 4\}$.

After training, class predictions were estimated for all periodically expressed oligonucleotides (identified using the methods described in Section 2.2) that were not included in the training data. We assigned the periodically expressed oligonucleotides to stage y if the maximum probability $\hat{p}_y$ was greater or equal to 0.8.

2.4.   Clustering of Periodically Expressed Genes on Chromosomes

From this point onwards, we worked with genes rather than with oligonucleotides. We used the Plasmodium Genome Resource (www.PlasmoDB. org) to obtain physical locations and ordering of all genes, and marked the stage assigned to each gene (if any). When different oligonucleotides from the same gene were assigned to more than one stage, we assigned the gene to the stage with the highest confidence estimate $\hat{p}_y$. Then we examined the patterns of periodically-expressed, stage-assigned genes along the 14 chromosomes. Using the chromosomal positions obtained, we defined a cluster as two or more consecutive loci whose expression patterns were matched to the same stage. Based on this definition, we could identify chromosomal clusters for each stage for a given cluster size. Figure 1 visually shows how the chromosomal clusters are defined based on the patterns of PEGs for a fictitious chromosome. On this fictitious chromosome, there are 30 genes of which 20 are periodically expressed. Solid blue, yellow, green and red colors represent PEGs assigned to stages 1–4 (ring/early trophozoite, trophozoite/early schizont, schizont and early ring stages), solid black symbols represents genes that are periodically expressed but were not assigned to a particular stage and open circles are genes that were not periodically expressed. It can be seen that there is one cluster of size 3 in stage "blue" and one yellow cluster of size 2 for the PEGs on this fictitious chromosome.

Original Data

Figure 1. Patterns of stage-specific periodically expressed genes in relation to chromosomal location. (See text for details.)

Original Data

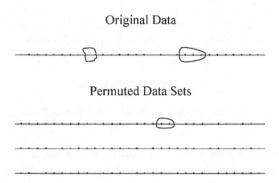

Permuted Data Sets

Figure 2. Assessment of significance of chromosomal clustering. In the original data there is one cluster of size 3 in stage "blue" and one yellow cluster of size 2. Three sample permutations are shown above, and one of the permutations gives a blue cluster of size 2.

In order to evaluate the statistical significance of stage-specific chromosomal clusters identified by previous algorithm, we propose a simple permutation test for this purpose. For each chromosome, we first randomly permuted the order of all the genes, but kept the assignment of periodicity and staging "attached" to each gene, then we counted the number of stage-specific clusters observed in the permuted datasets with a given cluster size. Figure 2 illustrates how the permutations were performed.

For a given number of permutations, B, we can calculate permutation p-values p^*_{smn} for a given cluster size s, chromosome m and stage n as

$$p^*_{smn} = \sum_{b=1}^{B} I\left(N^b_{smn} \geqslant N_{smn}\right)/B, \tag{2}$$

where $I(\cdot)$ is the indicator function, which equals 1 if the condition in parentheses is true and 0 otherwise. N^b_{smn} is the number of clusters on chromosome m and in stage n with cluster size s in the permutated data b. N_{smn} is the number of clusters on chromosome m and stage n with cluster size s in the original

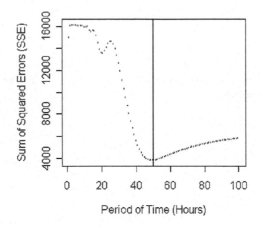

Figure 3. The relationship between the SSE and period.

data. We also calculated the statistical significance associated with the number of clusters on a particular chromosome, for a given cluster size, as

$$p_{sm}^* = \sum_{b=1}^{B} I\left(\sum_{n=1}^{4} N_{smn}^b \geqslant \sum_{n=1}^{4} N_{smn} \right) \Big/ B. \tag{3}$$

Equation (3) does not require matching the number of clusters assigned to each stage. For example, suppose the original data contained two clusters in stage 1 and one cluster in stage 3, for cluster size s on chromosome m. A permutation containing three clusters in stage 2 would be considered to have as many clusters as the original data. The permutation p-value p_s^*, counting the number of stage-defined clusters of a given cluster size across all chromosomes and stages is given as

$$p_s^* = \sum_{b=1}^{B} I\left(\sum_{m=1}^{14} \sum_{n=1}^{4} N_{smn}^b \geqslant \sum_{m=1}^{14} \sum_{n=1}^{4} N_{smn} \right) \Big/ B. \tag{4}$$

Similar to Equation (3), Equation (4) does not require exact matching of the number of clusters of each stage.

3.   RESULTS

3.1.   Estimation of the Cycle of Periodically Expressed Genes

We used the 472 oligonucleotides (351 genes) whose staging is known to estimate the period T by fitting Equation (1). Bozdech et al. [4] found that the

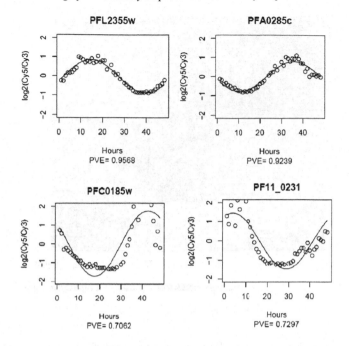

Figure 4. Examples expression profiles for 4 genes shown with a least-square fit of the data (curved line).

majority of gene profiles exhibited an overall expression period of 0.75–1.5 cycles per 48 h. For this reason we fitted Equation (1) over a range of 100 T values evenly spaced from 1 hour to 100 hours. As can be seen in Figure 3, the sum of squared errors over the 351 genes was minimized at 50 hours. Therefore, we selected $\widehat{T} = 50$ for subsequent analysis.

3.2.   Identification of Periodically Expressed Oligonucleotides

After fitting Equation (1), there were 2949 oligonucleotides (2204 genes) which showed evidence for periodic expression with PVE $\geqslant 0.7$ (F-statistic $\geqslant$ 50.2). Figure 4 shows examples of expression profiles for 4 genes, PFL2355w, PFA0285c, PFC0185w and PF11_0231. These genes were selected because they represent four distinct sine–cosine wave profiles in the dataset. The first peaks of the sine–cosine wave forms of these four genes were about 15 hours, 36 hours, 43 hours and 5 hours, respectively.

We observed that most of the genes which passed the PVE filtering criteria had one of these four profiles. This suggested that there were four dominant expression patterns in the selected periodically expressed genes.

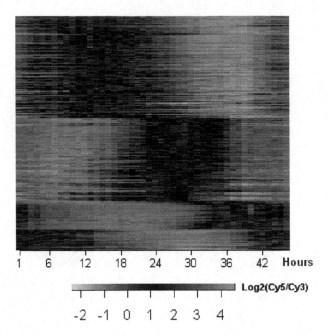

Figure 5. Heat map of periodically expressed genes predicted in four stages of IDC.

10,000 permutations of the data over the time points resulted in an estimated FDR was only 0.00003, based on the F-statistic cutoff of 50.2 (PVE $\geqslant$ 0.7), strongly suggesting that the randomized datasets do not demonstrate periodicity.

3.3.   Classification of Stage Group for Periodically Expressed Oligonucleotides

As previously noted, there are 472 oligonucleotides (351 genes) whose staging was known. These were used as the training samples in the SVM. Excluding these oligonucleotides, we had 2545 oligonucleotides (1918 genes) for testing. (It should be noted that some of the oligonucleotides in the training sample had PVE values less than 0.7, which explains why the number of oligonucleotides in the combined training and testing samples does not equal the number of periodically expressed oligonucleotides selected.) The pairwise binary SVM classifiers with the RBF kernel generated the 6 pairwise predictors. The 10-fold cross-validation error was 3.4%. For the 2545 oligonucleotides (1918 genes) of unknown stage, we assigned 923 oligonucleotides (718 genes) into ring/early trophozoite stage, 835 oligonucleotides (624 genes) into trophozoite/early schizont stage, 186 oligonucleotides (141 genes) into

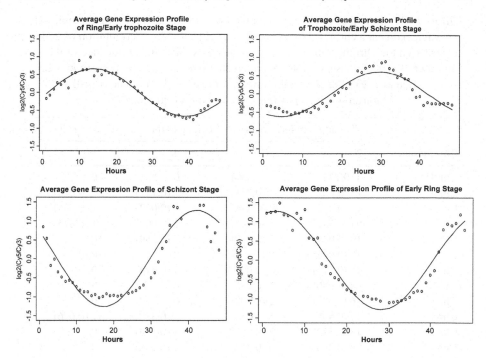

Figure 6. Meta-gene expression profiles of 4 stages. Sine–cosine curves were then fitted to the meta-expression profiles using Equation (1).

schizont stage and 199 oligonucleotides (167 genes) into early ring stage, each with an estimated class probability $\hat{p}_y$ of at least 0.8. Another 402 oligonucleotides (268 genes) that had class probabilities less than 0.8 were not assigned into any of these four stages.

Figure 5 shows the stageogram of the IDC transcriptome based on the 2143 classified oligonucleotides (1650 classified genes) which had class probability at least 0.8 and the 472 oligonucleotides (351 genes) in training set for which stage was known (class probability 1). First, the oligonucleotides were ordered by predicted stage; from top to bottom the ordering is ring/early trophozoite, trophozoite/early schizont, schizont and early ring, respectively. Secondly, within each stage, oligonucleotides were sorted by probability in descending order.

As can be seen in Figure 6, meta-expression profiles of each stage, calculated by averaging the expression values of all oligonucleotides predicted to be in the same stage over the 46 time points, are very similar to the profiles of the 4 representative genes shown in Figure 4. Our proposed method clearly identifies stage-specific patterns.

Table 1. Total number of clusters, summed over the four stages, (and empirical
p-values) on each chromosome, by different cluster sizes. Permutation of gene
order on the chromosomes was used to assess statistical significance of the
number of clusters observed of a given cluster size, on each chromosome using
Equation (3). For the total number of clusters observed across all
chromosomes, *p*-values are based on Equation (4)

Chromosome	# of adjacent loci predicted to belong to the same stage in a cluster			
	2	3	4	5
Chr-1	4 (0.133)	1 (0.136)	0	0
Chr-2	15 (0.0014)	2 (0.231)	1 (0.117)	1 (0.018)
Chr-3	14 (0.060)	2 (0.434)	2 (0.022)	1 (0.038)
Chr-4	9 (0.238)	3 (0.044)	2 (0.0037)	1 (0.014)
Chr-5	19 (0.034)	1 (0.870)	1 (0.259)	0
Chr-6	13 (0.0025)	0	0	0
Chr-7	15 (0.016)	5 (0.0026)	1 (0.114)	0
Chr-8	14 (0.080)	1 (0.732)	0	0
Chr-9	16 (0.159)	2 (0.509)	0	0
Chr-10	12 (0.818)	5 (0.064)	1 (0.319)	1 (0.076)
Chr-11	13 (0.579)	7 (0.0005)	1 (0.194)	0
Chr-12	18 (0.330)	3 (0.347)	1 (0.248)	0
Chr-13	33 (0.150)	15 (<0.0001)	2 (0.180)	0
Chr-14	43 (0.0173)	8 (0.167)	3 (0.067)	
Total	238 (0.066)	55 (0.091)	15 (0.106)	4 (0.040)

Total number of clusters: 312

3.4.　Chromosomal Clustering

In the remaining analysis, we focused on the 351 genes with known staging,
together with the 1650 genes whose estimated class probabilities were at least
0.8, for a total of 2001 genes. There were 990 genes which were measured
by more than one oligonucleotide. Table 1 shows the number of clusters on
each chromosome of different cluster sizes. A total of 238 clusters containing
2 loci, 55 clusters containing 3 loci, 15 clusters containing 4 loci and 4 clusters
containing 5 loci were identified. It should be noted that since the chromosomal
clusters were defined in a stage dependent way, the number of clusters for
each chromosome and cluster size in Table 1 is the total number of clusters
over all four IDC stages. For example, on chromosome 1 for cluster size 2, we
identified 2 clusters at trophozoite/early schizont stage, 1 cluster at schizont
stage and 1 cluster at the early ring stage, so the total number of clusters on
this chromosome is 4.

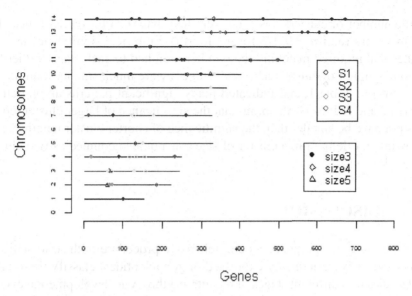

Figure 7. Whole chromosome view of 74 large clusters distributed on 14 chromosomes.

Figure 7 shows a whole genome view of the 74 larger clusters (where 3 or more adjacent genes were mapped to the same stage). Blue, yellow, green and red colors represent clusters identified at ring/early trophozoite, trophozoite/early schizont, schizont and early ring stages, respectively; circle, diamond and triangle symbols denote cluster sizes from 3 to 5, respectively. It can be seen that most large clusters were identified at ring/early trophozoite and trophozoite/early schizont stages with cluster size 3.

Our permutation analysis gave empirical p-values for each cluster size, within each stage and chromosome. We used 20,000 permutations to generate the empirical p-values. For these stage-specific results (Equation (2)), particularly small permutation p-values were associated with one stage 4 cluster of size 4 on chromosome 2 ($p = 0.00045$), two stage 3 clusters of size 3 on chromosome 2 ($p = 0.0002$), and two stage 4 clusters of size 3 on chromosome 11 ($p = 0.00065$). Since p-values were estimated for 14 chromosomes, 4 stages, and 4 cluster sizes (224 tests), these values should be interpreted cautiously and adjusted for multiple testing; a Bonferroni adjustment for $p = 0.05$ would consider only $p = 0.0002$ as significant. Therefore, we used the permutation analysis to also obtain summary p-values for the total number of clusters over all stages, and over all chromosomes. The total number of stage-specific clusters of size five (Equation (4)) was greater than the number expected by chance (Table 1; $p = 0.040$); for smaller clusters, the number of observed clusters also appeared to be slightly larger than expected (Table 1; size 2: $p = 0.066$, size 3: $p = 0.091$, size 4: $p = 0.106$). For the four chromosomes where clusters of size 5 were observed, chromosome-specific estimated p-values (Equation (3)) for the number of clusters observed were less than 0.04 for three of these four chromosomes (and $p = 0.076$ for the fourth). Empirical significance level estimates that required matching the number assigned to each stage (rather than the total number of clusters across all stages) were much smaller than the values shown in Table 1, and indicated many significant patterns among cluster sizes of 2 and 3. It is worth noting that the significance of larger cluster lengths does not have be smaller than the significance of shorter cluster lengths, since, following our definition, a cluster of size 5 is not also counted as two clusters of size 4.

4.   DISCUSSION

In this study we proposed a comprehensive procedure with solid statistical basis to identify periodically expressed oligonucleotides, classify these oligonucleotides into different stages of the intraerythrocytic developmental cycle of *P. falciparum* and map them to chromosomes to detect chromosomal clusters.

We identified 2949 oligonucleotides (2204 genes) were periodically expressed in *Plasmodium falciparum* by our definition, suggesting that almost half of the *Plasmodium falciparum* genes (2204 out of 4488) are transcriptionally involved in stage-specific activities. For comparison, Spellman et al. [16] identified only 800 PEGs out of 6178 yeast genes. Most of the PEGs were assigned to either the ring/early trophozoite or the trophozoite/early schizont stages of the IDC. Our IDC stageogram demonstrates clear boundaries among the four IDC stages, unlike Bozdech's study [4] where the stageogram showed a cascade of continuous expression. Due to our selection criteria of (a) at least 70% of variation explained by the periodicity, and (b) stage classification probability of at least 80% the oligonucleotides in our stageogram were highly selected for clear and consistent periodic signatures.

We identified many more clusters than Bozdech et al.'s study [4]. They defined a chromosomal cluster as a region in which the correlation of 70% of the possible pairs of adjacent genes on the same chromosome was greater than or equal to 0.75. Based on this criterion, they found only 37 clusters consisting of 3 genes and 14 clusters consisting of more than 3 genes. In our study, there were 55 clusters with 3 genes and 19 clusters consisting of more than 3 genes. Many clusters detected in their study were also found in our study. For example, 34 of 51 larger clusters (3 genes or larger) identified in their study were also found in the 74 larger clusters we detected. The seven genes of the SERA family that they found on chromosome 2 [13] were observed in two of our clusters. The first SERA gene cluster contained two genes in the trophozoite/early schizont stage, and a second SERA gene cluster contained 5 genes in the schizont stage. Based on our study and that of Bozdech et al. [4], it appears that there were few large clusters in the *P. falciparum* genome. Most (94%) of the chromosomal clusters that we identified were of size 2 or 3. It is also interesting to note that there was no obvious difference in cluster-distribution across the chromosomes; for example, approximately 33% of the clusters were on the two longest chromosomes 13 and 14, and these chromosomes form approximately 35% of the total genome length.

In addition, we downloaded gene annotations with GO terms and EC for *P. falciparum* strain 3D7 from www.PlasmoDB.org. Our primary analysis showed that some PEGs with similar functions are clustered together. For the larger clusters, where there are 3 or more adjacent genes in a cluster, Bozdech et al.'s study [4] found only two clusters (SERA gene cluster and ribosomal protein gene cluster) out of the 51 large clusters where the genes were known to have a functional relationship. However, we found 11 clusters (including the above two) out of our 74 larger clusters that contained at least two loci whose annotation clearly indicated that the genes are functionally related. For example, we identified an energy gene cluster (PF10_0121, PF10_0122 and PF10_0123) assigned to the ring/early trophozoite stage on chromosome 10. An RNA processing gene cluster (MAL13P1.322, MAL13P1.323 and PF13_0340) and an ATP binding gene cluster (PF13_0177, PF13_0178, PF13_0179 and PF13_0180), both assigned also to the ring/early trophozoite stage, were found on chromosome 13. The permutation *p*-values for these three clusters are 0.275, 0.0018 and 0.045, respectively. This information may be useful when annotating the function of the many unknown gene products in the *P. falciparum* genome.

It should be noted that there are some limitations in this analysis. The choice of a single value for the period T for all oligonucleotides may not be optimal, and therefore there may be additional PEGs with shorter or longer cycles. Furthermore, the sinusoidal model in Equation (1) will not do a good job of identifying "spikes", or genes whose action is of short duration that may initiate stage-specific patterns of expression. Therefore, such genes will be left out of the identified clusters. Another concern is that our estimate of the FDR for identifying periodically expressed oligonucleotides was very small, which gives rise to concern about underestimation. One possible reason for a downward bias in FDR is that there were significant serial correlations in the expression levels of a given gene over time due to the slowly varying nature of the cell culture. Anderson et al. [1] pointed out that permutation of raw data under the full model will not maintain type I error close to a nominal α when there is collinearity among the independent variables. They suggested that permutation of residuals under a serial correlation model would be a better choice in this case.

The definition of a gene "cluster" is inherently somewhat ad hoc. Our definition is quite restrictive, since only adjacent PEGs with the same assigned stage are called clusters. In fact, some of these clusters may represent tandem gene duplications. Our definition leaves out genes whose periodicity may be harder to detect, clusters made up of genes acting in different stages but in the same functional capacity, or clusters where there are one or more unrelated genes lying within the cluster. Following the latter idea, clusters could be defined by estimating the density, along the chromosome, of genes that are predicted to be related [6]. Nevertheless, our approach identified several clusters that seemed to be functionally important when gene function was examined.

Our permutation analysis gave empirical significance levels associated with each stage, each cluster size, and each chromosome. We also estimated, for a given cluster size, the significance levels for the total number of clusters in any stage on each chromosome, and the number across all chromosomes, and these empirical p-values are intrinsically adjusted for the multiple testing involved. The results suggest a small excess of clusters over the number that might be expected by chance, but in fact we can expect these results to reflect a conservative estimate of the number of real stage-specific clusters due to the imperfect prediction of PEGs and stages, and the strictness of our definition of a cluster. Hence, our analysis provides evidence for stage-specific cis-regulation within functional clusters.

REFERENCES

[1] Anderson, M.J. and Legender, P., An empirical comparison of permutation methods for tests of partial regression coefficients in a linear model, *Journal of Statistical Computation and Simulation*, **62** (1999), 271–303.

[2] Benjamini, Y. and Hochberg, Y., Controlling the false discovery rate: A practical and powerful approach to multiple testing, *Journal of the Royal Statistical Society B*, **85** (1995), 289–300.

[3] Booth, J.G., Casella, G., Cooke, J.E.K., and Davis, J.M., Clustering periodically expressed genes using mciroarray data: A statistical analysis of the yeast cell cycle data, Statistics Department Technical Report, University of Florida, 2003.

[4] Bozdech, et al., The transcriptome of the intraerythrocytic developmental cycle of *Plasmodium falciparum*, *PloS Biology*, **1** (2003), 1–16.

[5] Caron, H. et al., The human transcriptome map: Clustering of highly expressed genes in chromosomal domains, *Science*, **291** (2001), 1289–1292.

[6] Chaudhuri, P. and Marro, J.S., SiZer for exploration of structures in curves, *Journal of the American Statistical Association*, **94** (1999), 807–823.

[7] Cohen, B.A., Mitra, R.D., Hughes, J.D., and Church, G.M., A computational analysis of whole-genome expression data reveals chromosomal domains of gene expression, *Nature Genetics*, **26** (2000), 183–186.

[8] Dudoit, S., Shaffer, J.P., and Boldrick, J.C., Multiple hypothesis testing in microarray experiments, *Statistical Science*, **18** (2003), 71–103.

[9] Florens, L. et al., A proteomic view of the *Plasmodium falciparum* life cycle, *Nature*, **419** (2002), 520–526.

[10] Friedman, F., Another approach to polychotomous classification, Statistics Department Technical Report, Stanford University, 1996.

[11] Hastie, T. and Tibshirani, R., Classification by pairwise coupling, *The Annals of Statistics*, **26** (1998), 451–471.

[12] Lu, X., Zhang, W., Qin, Z.S., Kwast, K.E., and Liu, J.S., Statistical resynchronization and Bayesian detection of periodically expressed genes, *Nucleic Acids Research*, **32** (2004), 447–455.

[13] Miller, S.K. et al., A subset of *Plasmodium falciparum* SERA genes are expressed and appear to play an important role in the erythrocytic cycle, *Journal of Biology Chemistry*, **277** (2002), 47524–47532.

[14] Platt, J., Probabilistic outputs for support vector machines and comparison to regularized likelihood methods, in: A. Smola, P. Bartlett, B. Schoelkopf, and D. Schuurmans (Eds.), *Advances in Large Margin Classifiers*, MIT Press, Cambridge, MA, 2000.

[15] Roy, P.J. et al., Chromosomal clustering of muscle-expressed genes in *Caenorhabditis* elegans, *Nature*, **418** (2002), 975–979.

[16] Spellman, P.T. et al., Comprehensive identification of cell-cycle-regulated genes of the Yeast saccharomyces cerevisiae by microarray hybridization, *Molecular Biology of the Cell*, **9** (1998), 3723–3297.

[17] Taylor, J., Tibshirani, R., and Efron, B., The "Miss rate" for the analysis of gene expression data, *Biostatistics*, **6** (2005), 111–117.

[18] Troyanskaya, O. et al., Missing value estimation methods for DNA microarrays, *Bioinformatics*, **17** (2001), 520–525.

[19] Whitfield, M.L. et al., Identification of genes periodically expressed in the human cell cycle and their expression in tumors, *Molecular Biology of the Cell*, **13** (2002), 1977–2000.

[20] Vapnik, V., *Statistical Learning Theory*, Wiley, 1998.

Chapter 9

PlasmoTFBM: An Intelligent Queriable Database for Predicted Transcription Factor Binding Motifs in *Plasmodium falciparum*

Chengyong Yang[a], Erliang Zeng[a], Kalai Mathee[b], Giri Narasimhan[a,*]

[a]*Bioinformatics Research Group (BioRG), School of Computing and Informations Sciences, Florida International University, Miami, FL 33199, USA*
[b]*Department of Biological Sciences, Florida International University, Miami, FL 33199, USA*

Abstract   There is very little information available with regard to gene regulatory circuitries in *Plasmodium falciparum*. In an attempt to discover transcription factor binding motifs (TFBMs) in *P. falciparum*, we considered two approaches. In the first approach, gene expression data from asexual intraerythrocytic developmental cycle generated every hour for 48 hour post-infection were fed into the ISA (Iterative Signature Algorithm), which outputs modules composed of sets of genes associated with co-regulating conditions. Putative TFBMs were discovered by applying the AlignACE program on the resulting gene sets. In the second approach, the MotifRegressor program was used to predict potential motifs associated with induced and repressed genes for each time point and then clustered based on the strength of their correlation to the gene expression (i.e., motif coefficients) across different time points. A total of 637 and 840 putative motifs were predicted by the MotifRegressor and ISA-AlignACE programs, respectively. All this information was uploaded into a database, thus making it easy to devise complex queries. Using published information on known motifs, we were able to validate some of our results. In addition, modules consisting of putative transcription factors and related genes were also investigated. This work provides a bioinformatics methodology to analyze transcription regulation and TFBMs across the whole genome. By constructing a comprehensive relational database and an intelligent, user-friendly query system, biologically meaningful conclusions can be drawn easily even by an investigator with no prior knowledge of databases.

Keywords: transcription factor, regulatory elements, motifs, G-box, *SPE* elements, *CPE* elements, *var* genes, heat shock protein, SERA, EBA140, *Plasmodium falciparum*

---

*Corresponding author.

1.   INTRODUCTION

The challenge of CAMDA'04 was to analyze the gene expression data, which was generated by DeRisi's laboratory using transcripts from the organism *Plasmodium falciparum*, harvested at 46 different time points during its intraerythrocytic developmental life cycle (Bozdech et al., 2003). *P. falciparum* is one of four species of the parasitic protozoan *genus Plasmodium*, and is responsible for the vast majority of malaria episodes, affecting 200–300 million individuals and causing 0.7–2.7 million deaths per year worldwide (http://www.who.int/malaria).

In this paper, we focused on mining for information related to gene regulation and transcription factor binding motifs (TFBM), which is important considering the fact that direct experimental identification of TFBMs is slow and laborious. We used two recently developed algorithms to predict potential TFBMs: AlignACE (Hughes et al., 2000; Roth et al., 1998) and MotifRegressor (Conlon et al., 2003; Liu et al., 2002). Using the limited information on known motifs, we were able to validate some of our results.

The AlignACE (**Align**s **N**ucleic **A**cid **C**onserved **E**lements) program is best applied on sets of co-regulated genes. Standard clustering tools such as hierarchical, K-means clustering, and self-organizing maps assign genes to unique clusters by relying on the similarity of the expression profiles of the co-regulated genes across all conditions for their identification (Han and Kamber, 2001). However, many genes play multiple roles under various conditions in complex, interrelated biological processes. We, therefore, obtained clusters of potentially co-regulated genes by using the Iterative Signature Algorithm (ISA), which (a) allows for clustering of genes that exhibit similarity of the expression profiles only at specific sets of time points, and (b) allows for genes to be part of multiple clusters (Ihmels et al., 2002). This permits the ISA approach to explore complex interrelationships among genes. It outputs a set of transcription modules, each of which is a self-consistent unit consisting of potentially co-regulated genes and the regulating conditions (Ihmels et al., 2004).

One of the difficulties with the motif discovery programs is that they produce a large number of predicted TFBMs along with associated scores representing the statistical significance of the predictions. However, drawing biologically useful inferences or conjectures remains a difficult problem. In this paper, we present a new approach that will facilitate the process of drawing meaningful conclusions that are likely to be useful to a biologist. This is achieved by constructing a comprehensive relational database for *Plasmodium falciparum* with the predicted Transcription Factor Binding Motifs called **PlasmoTFBM** (Figure 1), and an intelligent, user-friendly query system.

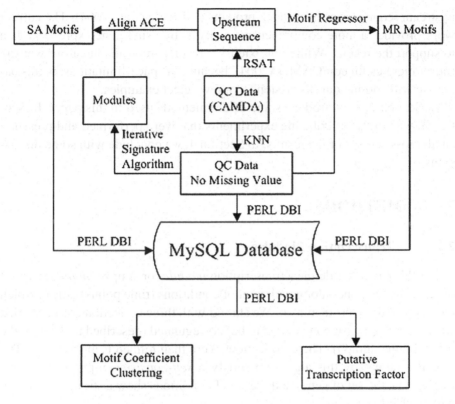

Figure 1. Flowchart for mining TFBMs for *P. falciparum*.

The PlasmoTFBM database contains the following information:

1. All the discovered TFBMs, along with their significance scores, the software using which they were found, and the genes whose upstream sequences contained them along with their location in those upstream sequences.
2. Clusters of co-regulated genes (referred to as transcription modules, or simply modules), and the time points at which they were found to be co-regulated.
3. All genes and ORFs in the genome, their chromosomal location, their functional annotation, and their expression information at all the time points during the development of the parasite.

We show, with examples, how an investigator can generate "conjectures" using this database, which could then be used to perform directed laboratory experimentation.

The only other related work on studying genome-wide TFBMs in *P. falciparum* is by Militello et al., where they applied the AlignACE software to the upstream sequences of heat shock proteins (Militello et al., 2004). The current work provides a more comprehensive analysis by using gene expression data to support the results. While our extensive results are available at our website (http://biorg.cs.fiu.edu/CAMDA2004), because of space-limitations, in this paper we will confine our discussions to a few select examples.

In Section 2, we introduce some of the methods used in this paper. In Section 3, we briefly describe the experiments that were performed and present a small cross-section of the results. In Section 4, we conclude with some discussions.

2.   METHODS

2.1.   Transcription Modules

For this paper, we define a transcription *module* (or simply, *module*) as a set of co-regulated genes along with a set of conditions (time points) during which they appear to be co-regulated. We started with three collections of genes that were known to be (or conjectured to be) co-regulated (described in detail in the following paragraph). These collections were then refined using the ISA. The modules output by this algorithm satisfy a *self-consistency* property, which implies that the set of genes and the set of conditions show a strong correlation with each other.

Transcription modules were generated in several different ways, each time by applying the ISA algorithm (Bergmann et al., 2003; Ihmels et al., 2002). A first set was generated by starting from a specific interesting gene. For this paper, 13 putative transcription factors were chosen. They are MAL13P1.213, MAL7P1.86, MAL8P1.131, PF07_0057, PF10_0143, PF13_0043, PF14_0469, PFA0525w, PFB0290c, PFB0730w, PFE0305w, PFE0415w, and PFI1260c. A second set of modules was generated by starting from collections of genes known to be involved in the same function (e.g., heat shock proteins); such sets were obtained from the PlasmoDB website (http://www.plasmodb.org) (Bahl et al., 2003). A third set was generated by starting from random initial sets. User-defined thresholds for the ISA method were chosen as follows: gene thresholds were selected from 1.0 to 2.5 with a step of 0.1, and condition threshold was fixed at 2 (it was held constant because its choice had a negligible effect on the output over a comparable range, as was also observed in Ihmels et al., 2004). In total 217 transcription modules were obtained with gene sets ranging in size from 10 to 500. All the 217 modules can be found on the supplemental website at (http://biorg.cs.fiu.edu/CAMDA2004/).

2.2. AlignACE

AlignACE is a Gibbs sampling algorithm for detecting motifs that are over-represented in a set of DNA sequences (Hughes et al., 2000; Roth et al., 1998). A C++ implementation was downloaded from their website [http://atlas.med.harvard.edu]. The upstream sequences of co-regulated genes obtained from the transcription modules (described above) were downloaded, and AlignACE was used to search for motifs in them. For our experiments with AlignACE, the GC content was set at 19.36%, the GC-content of the *P. falciparum* genome (Gardner et al., 2002).

2.3. MotifRegressor

MotifRegressor is a second motif-detection tool used in this work. It first uses MDscan as a feature extraction tool to construct candidate motif matrices and then applies regression analysis to select motifs that are strongly correlated with changes in gene expression (Conlon et al., 2003; Liu et al., 2002). For our experiments, the upstream sequences were cleaned up so that single repeats of at least 10 bases and double repeats of at least 16 bases were removed. As mentioned below, MotifRegressor was applied separately on gene expression data from the 46 time points. MotifRegressor has the advantage of using a more sophisticated background model (third-order Markov model), and selects for motifs that explain the data and correlate with the expression behavior of interest. It also provides significance scores for the discovered motifs.

For most part, we used the default settings for MotifRegressor. In this procedure, upstream sequences are first ordered by their relative gene expression values, then the top 50 sequences are chosen as a seed to obtain matrices for w-mer motifs (here we used $5 \leqslant w \leqslant 15$). Using a semi-Bayesian scoring function, the 50 highest-scoring motifs are obtained and then refined by using the 250 sequences with the highest relative gene expression values. Sequence *Motif-Matching Score* is generated in this step to determine how well the upstream sequence of a gene g matches a motif m. For motifs reported by MDscan, gene expression values were regressed on sequence motif matching score using a stepwise linear regression procedure. The candidate motifs with a significant p value ($p \leqslant 0.01$) are retained.

2.4. Data

The gene expression data that passed all quality control filters (QC data) were downloaded form the CAMDA website. The gene expression data was available for every hour up to 46 hours post-infection (hpi). Standard R package routines (based on the K nearest neighbor method) were used to impute missing values (Troyanskaya et al., 2001). Regulatory Sequence Analysis Tools were used to extract upstream sequences for the ORFs (van Helden,

2003). For the analysis, the length of the upstream sequences used was 2000 bp.

2.5.   Generating Potential TFBMs

The QC data and the corresponding upstream sequences were analyzed. The ISA algorithm was applied on available collections of related genes. The resulting transcription modules were used as initial sets to run AlignACE resulting in one set of motifs. Then, the MotifRegressor software was ran on the gene expression data for each of the 46 time points separately, to obtain 46 sets of significant motifs. Motifs with identical consensus sequences were merged using Perl scripts (the cleaning step). There were 1077 motifs generated from MotifRegressor and 936 from AlignACE. After the cleaning step, 637 MotifRegressor and 840 AlignACE motifs remained.

2.6.   Database

A relational database called PlasmoTFBM was designed and implemented using MySQL to store all the available information. This includes the gene expression data, generated significant motifs and modules, gene annotation information including the functional information and the chromosomal location. Figure 1 shows the scheme used for the analyses of the data.

2.7.   Web Query and Visualization

Web query interface was implemented using PHP (**PHP: H**ypertext **P**reprocessor). Although it is possible to design complex queries for the PlasmoTFBM database using Perl DBI, it requires non-trivial expertise to be able to use it effectively. The motivation for the query system was as follows. Most biologists perform research on a small set of genes, usually a set of genes that are involved in a specific function or a specific pathway. Such a biologist would be interested in knowing whether this database has results that are relevant to their genes of interest, i.e., what other genes are co-regulated with the ones in questions, what motifs might they share, what developmental stage or functional pathway might they be involved in, what transcriptional factors may be regulating the genes of interest, and finally, what biologically meaningful conjectures can result from the analyses and that may be relevant to the genes of interest. Answers to such questions may be the starting point for further investigations for the biologist. A handy web-based query system could automate some of the analyses.

Consider the following example. Assume that the genes of interest are *MAL13P1.60* and *MAL7P1.86*. The *MAL13P1.60* encodes the protein erythrocyte-binding antigen 140 (EBA140), which is implicated in merozoite invasion using a sialic-acid-dependent receptor on human erythrocytes (Baum et al.,

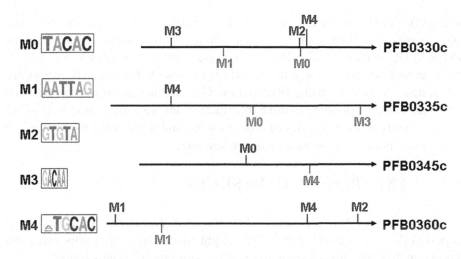

Figure 2. Motifs visualized in the upstream sequence of all the SERA genes of interest. The line indicates the upstream sequence with the translation start site at the right end. Motifs labeled as M0, M1, M2, M3, and M4 correspond to motifs Motif.P29.5.3BG, Motif.N29.6.15BG, Motif.P31.5.22BG, Motif.P33.5.10BG, and Motif.P35.6.3BG, respectively. Red color represents motifs in the forward direction while blue color represents those in the reverse direction.

2003; Bozdech et al., 2003; Thompson et al., 2001). The open reading frame, *MAL7P1.86*, codes for a putative alpha subunit of transcription initiation factor IIE (TF IIE). In addition, both genes are expressed highly during the merozoite stage of the parasite's development. A biologist may be interested in studying their relationship: Are they co-regulated (i.e., is there a transcription module that contains both of them, is there a set of time points or conditions under which their expression profiles are correlated)? Do they share any motifs in their upstream regions? Can any other relationships be conjectured?

In the first step, all transcription modules that include some subset of the genes of interest are computed, sorted by the number of genes of interest that they contain. Next, the user may choose a subset of the generated modules for further exploration. Suppose that the user decides to explore the module MAL7P1.86_g2_c6, which contains both the genes of interest. The query system also outputs the conditions (hpi 1, 27, and 41–45 in this example) and gene sets associated with the selected module. The user could then ask for the list of all the motifs found in the module MAL7P1.86_g2_c6, and under the selected conditions (say, hpi 27, 42 and 45 in this example).

Visualization tools are provided to visualize the final results in a more meaningful way. All motifs of interest are displayed using the WebLogo notation (Crooks et al., 2004). The user may select a specific set of genes, and the motifs for each gene of interest is then displayed in a graphical manner by showing their location as a function of their distance from the translation start site (ATG) in the upstream sequence of the gene. See Figure 2 for an example, where we looked at group of genes that code for serine-repeat antigens (SERA) (Rosenthal, 2004). The average gene expression profile along with the standard deviation is also displayed for the selected genes. All images are generated dynamically using PHP and the GD graphics library. Thus, the web interface makes it possible to mine information and visualize some interesting results simply through a series of mouse clicks, and without the user having to learn a complicated database or a query language.

3.   EXPERIMENTAL RESULTS

There are very few regulatory elements in *P. falciparum* that have been reported (Horrocks et al., 1998). We sought to validate our results using the known motifs. We discuss some of the interesting motif groups found.

3.1.   G-Box Motifs

Recently, a novel G-rich regulatory element named G-box was identified upstream of several *P. falciparum hsp* genes (Militello et al., 2004). Since the genome of *P. falciparum* is AT-rich (only 19.36% GC content), the G-box is considered a unique regulatory element. We investigated motifs in seven genes corresponding to heat-shock proteins (Hsp) or putative Hsps. The G-box was also found by our analyses in all these seven *hsp* genes (Figure 3). Furthermore, our analysis showed that the G-box motif was found to be significant at all 46 time points, and was not confined to just the *hsp* gene family, suggesting that the G-box is a common regulatory element, and is not stage-specific.

Next, we compared the motif sequences found by our analyses with the published sequence, (A/G)NGGGG(C/A) (Militello et al., 2004). However, the AlignACE method found several longer motifs containing the published sequence for G-box. The variants of these motifs found are shown in Figure 3.

3.2.   Motifs in *var* genes

It is known that there are nearly 50 diverse *var* genes distributed throughout the parasite genome coding for variants of *P. falciparum* erythrocyte membrane protein 1 (PfEMP1); they are responsible for both antigenic variation and cytoadherence of infected erythrocytes in malaria (Voss et al. 2000, 2003). The ability of the parasite to switch the expression of PfEMP1 allows it to escape specific immune responses, and changes in its antigenic phenotype correlate with the altered properties of PfEMP1 (Voss et al. 2000, 2003). Thus understanding the regulatory mechanisms of PfEMP1 variants and other genes is very critical.

Locus	hpi	WebLogo	Motif Score
PFI0875w (HSP)	26-34, 39-45		15.58
MAL8P1.143 (hypothetical)	1-48		113.62
PF08_0032 (hypothetical)	1-3, 6, 27-37, 41-48		38.32
PF11_0175 (HSP 101)	11-18, 26-33		15.28
PF11_0188 (HSP 90)	1-48		50.08
PF11_0351 (HSP 70)	1-48		222.77
PFL0740c (hypothetical)	1-48		132.64

Figure 3. G-box motifs appeared in the upstream sequences of the *hsp* genes given in column 1. The motifs shown using the WebLogo format (Crooks et al., 2004) were obtained by using AlignACE on modules that included the hpi mentioned in the second column. The AlignACE method provided the motif scores mentioned in the last column (Hughes et al., 2000).

It was observed previously that most of the *var* genes were expressed in the early ring stage, but only one *var* gene variant is induced in the trophozoite stage, while the others are silent. We queried our database to find the motifs contained in the *var* genes. Our analysis showed the presence of two significant motifs (Figure 4): one was observed in a cluster of *var* genes at hpi 11 associated with inducing effect, while another motif at hpi 38 associated with repressing effect.

Previous studies of *var* genes have shown that nuclear proteins bind to conserved sequence motifs called *SPE1* (CACGGACACATGCAGTAACCGA-GAATTATTATATATAAATAT) and *SPE2* (T**GTGCATA**GTGGTGCG) and *CPE* (ATGT**TGTACAT**) (Voss et al., 2003). These were found by transfection experiments, and not by the use of sequence analysis or motif prediction software (Voss et al., 2003).

We used the motif sequence information and queried our database. We found motifs in our database that were subsequences of the *SPE2* and *CPE* ele-

Locus	Stage	Motif effect	WebLogo	Motif Score
PFL0935c PF14_048 PFI1830c PF10_0406 PFL1955w PFA0765c PFD0615c PFB0010w PF08_0103	Ring	Induce	 TATGTAcATA	3720.20
PFD0230c PF08_010 PFL0935c PF10_040 PFB0010w PFI1830c PFA0765c	Schizont	Repress	 gTGCATA	4249.90

Figure 4. Some significant motifs from the *var* genes. The first one contains part of the *CPE* motif, while the second one contains a part of the *SPE2* motif (Voss et al., 2003). WebLogo was used to display the motif. The motif scores are the result of using MotifRegressor program (Conlon et al., 2003).

ments reported previously (Figure 4). The portions of *SPE2* and *CPE* that over-lapped with our motifs are underlined above. In addition, our analysis showed that similar motifs were significant in a group of *var* genes that were induced at the ring stage. In contrast, the extended *SPE2* element was found in a group of *var* genes that were repressed at the schizont stage. However, these motifs were not unique to the group of *var* genes, but were also present in other genes at the ring and schizont stages. The analysis of the *SPE1* sequence did not generate any potentially useful interpretations.

3.3.   Discovery of Multiple Motifs

The MotifRegressor program predicted a total of 637 significant motifs across the 46 time points. The motifs were then clustered by motif coef-ficients, as suggested by Conlon et al. (2003). In brief, for each motif, at each time point, the gene expression values were regressed against the up-stream sequence motif-matching scores (reported by the MDscan component of MotifRegressor). Consequently, each motif can be represented by a vec-tor of 46 simple regression coefficients. The 637 motifs were then hierarchi-cally clustered into 12 groups based on the Euclidean distances between their

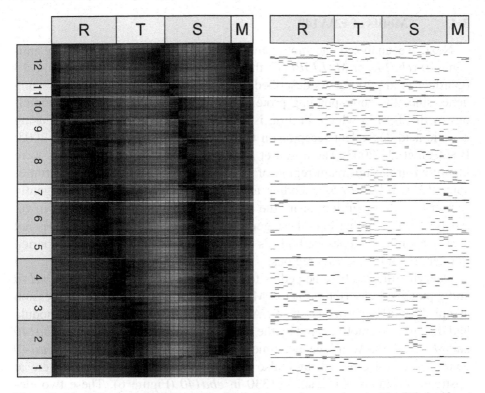

Figure 5. Motif clusters from cell cycle expression time series experiments. The 637 significant motifs reported by MotifRegressor over one cell cycle are clustered by motif coefficients over 46 time points. This figure was produced using Genesis software package by applying hierarchical clustering with Euclidean distance metric on the motif coefficient data (Sturn et al., 2002). Red shades correspond to positive motif coefficients (and, therefore positive correlations with the expression of the downstream genes), while green shades correspond to negative coefficients. The figures indicate the stages of the parasite (R – Ring, T – Trophozoite, S – Schizont, M – Merozoite) and the 12 clusters of motifs obtained.

coefficient vectors. The motif coefficients can be interpreted as the putative influence of a particular motif on the expression of downstream genes. Figure 5 shows the clusters of motifs with the plot on the left showing the motif coefficients across all time points. The plot on the right side shows the time points when the corresponding motifs were discovered as being significant. As can be seen in the figure, a majority of the motifs showed a periodic behavior, indicating that they are regulated periodically during the *P. falciparum* IDC. The above analysis showed that many motifs were found at the time points at which they were known to have the strongest effect (see supplemental material "Time point distribution of motif clusters" at the website [http://biorg.cs.fiu.edu/CAMDA2004/]).

3.4.   Motifs of EBA140

Next, we analyzed the motifs in the gene for erythrocyte-binding antigen 140 (*eba140* or *MAL13P1.60*) that lies in *P. falciparum* chromosome 13
(Gardner et al., 2002). As described before, this is a particularly interesting
gene, since the corresponding protein shares structural features and homol-
ogy with EBA175 which, in turn, is implicated in merozoite invasion using
a sialic-acid-dependent receptor on human erythrocytes (Baum et al., 2003;
Bozdech et al., 2003; Thompson et al., 2001). Eight significant motifs were
identified in the upstream region of *eba140*. The adjacent gene on chromosome 13 is *MAL13P1.61* encoding a hypothetical protein that is divergently
transcribed, and therefore share the upstream promoter region with *eba140*.
Analysis suggests that both these genes are tightly co-regulated, and it is not
clear which of the genes (or both) is regulated by the putative motifs reported
in their common upstream regions.

Querying the database helped us to locate a module that contained *eba140*
and a putative transcription factor, MAL7P1.86, which has a peak expression
at hpi 42 (early merozoite stage). AlignACE, when applied to this module
had discovered a motif shared by the upstream sequences of the genes *eba140*
and *MAL7P1.86*. At the spanned time period, this *MAL7P1.86* and the *eba140*
genes were co-expressed; they also shared common motifs, which were at
upstream locations −752 and −1330 in *eba140* (Figure 6). These two elements have very similar core sequence ("ACACA"). These two motifs were
also shared by 77 other genes that are highly expressed at 41 hpi. One possible
conjecture is that these genes are regulated by MAL7P1.86 by interacting with
these two TFBMs. This would then suggest that MAL7P1.86 is auto-regulated.
Alternatively, one could also conjecture that these genes are activated by an
unknown transcription factor that interacts at these motifs.

It is worth pointing out that the above analysis on *eba140* and *MAL7P1.86*
was easily performed as a sequence of straightforward queries of our database. Our belief is that with the help of domain-specific experts we can easily
generate more biologically meaningful conjectures using a database such as
PlasmoTFBM.

4.   DISCUSSION AND CONCLUSIONS

Using the ISA approach, transcription modules were generated. Each module consists of a set of potentially co-regulated genes along with a set of time
points at which the regulation is potentially occurring. Correlation and depen-
dencies between the conditions can be used to elucidate system-level transcriptional relationships. Compared to other existing clustering approaches (Eisen
et al., 1998; Tamayo et al., 1999), the ISA algorithm does not require the genes

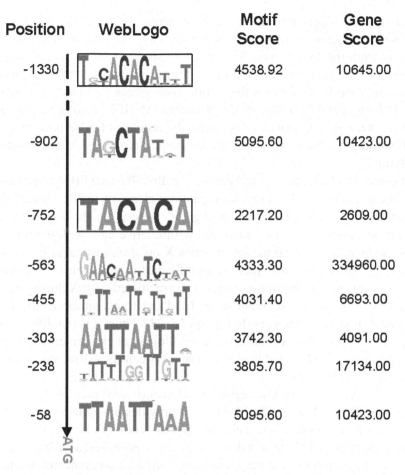

Position	WebLogo	Motif Score	Gene Score
-1330		4538.92	10645.00
-902		5095.60	10423.00
-752		2217.20	2609.00
-563		4333.30	334960.00
-455		4031.40	6693.00
-303		3742.30	4091.00
-238		3805.70	17134.00
-58		5095.60	10423.00

Figure 6. Motifs found in upstream of gene *eba140*. Boxed motifs are motifs shared by genes *eba140* and the divergently transcribed *MAL7P1.86* encoding a putative transcription factor (as well as other 77 other genes). Motif scores were as reported by the MotifRegressor program. The gene score shown on the last column indicates how well the upstream sequence of a gene matches a motif in terms of both degree of matching and number of sites (Conlon et al., 2003).

in a cluster to be correlated under all the conditions. It also allows genes to be part of multiple modules, which is a likely event since many genes are involved in different pathways at different time points.

We applied two existing motif detection tools on the CAMDA data sets. Both methods found a large number of potential transcription factor binding motifs. Our results on the G-box motifs support the conclusion that this organism may have unique regulatory mechanisms different from other known eukaryotic organisms (Militello et al., 2004). By design, the two approaches will find sequence motifs that are enriched in the input sequences (AlignACE) or best match the expression pattern (MotifRegressor). However, false positives are inevitable. AlignACE, in particular, is prone to give high scores to over-represented sequences in low-complexity regions, even though more stringent clusters from the ISA approach were used. Mechanisms to remove spurious results are extremely critical, but difficult and are themselves error-prone. In this current work, we rely on the significance scores provided by AlignACE (MAP Scores) and MotifRegressor (Sequence Motif-Matching Score) to provide the necessary guidance to decrease the number of false positives. More sophisticated mechanisms to improve the quality of the results are planned for the future.

We have implemented a novel database called PlasmoTFBM containing information relating to *P. falciparum* regulatory elements in IDC, which can be a useful tool to facilitate further biological research on the organism. Some sample questions that can be answered with relative ease with the use of our database include: (a) Find the set of genes X on chromosome A between loci L_1 and L_2. (b) Find motifs that are significant for set X during the schizont stage. (c) Locate a transcription factor Y co-regulated with X during the early merozoite stage or late schizont stage. (d) Does transcription factor Y share any motifs that are significant during hpi 18–21? Thus, it is possible to "bootstrap" any information available from the biological experiments to generate new and useful (and plausible) conjectures that can then drive future directed laboratory experiments.

Considering that very few regulatory elements were previously known for *P. falciparum*, the PlasmoTFBM database provides a useful pool of potential targets for investigators. It is well known that genes can be regulated both at the transcriptional and the translational stages. Recent research has suggested that the post-transcriptional gene regulation may be a predominant mechanism used by *P. falciparum* (Coulson et al., 2004; Hall et al., 2005). However, this does not diminish the importance of transcriptional regulation. Thus our database could still play an important role in revealing the putative regulatory elements involved in the transcriptional stage.

We provide a website (http://biorg.cs.fiu.edu/CAMDA2004), which will contain all the motifs and modules discovered by our analyses. We also provide a website (http://biorg.cs.fiu.edu/TFBM/) for web query and data visualization.

ACKNOWLEDGEMENTS

Authors C.Y. and E.Z. contributed equally to this paper. We thank Dr. X. Shirley Liu for help with MotifRegressor. We thank Gaolin Zheng and Haifeng Wang for helpful discussions. E.Z. is supported by a Florida International University Presidential Graduate Fellowship. G.N. is supported in part by NIH Grant P01 DA15027-01.

REFERENCES

Bahl, A., Brunk, B., Crabtree, J., Fraunholz, M.J., Gajria, B., Grant, G.R., Ginsburg, H., Gupta, D., Kissinger, J.C., Labo, P., Li, L., Mailman, M.D., Milgram, A.J., Pearson, D.S., Roos, D.S., Schug, J., Stoeckert, Jr., C.J., and Whetzel, P. (2003), PlasmoDB: The *Plasmodium* genome resource. A database integrating experimental and computational data, *Nucleic Acids Res.*, **31**(1), 212–215.

Baum, J., Thomas, A.W., and Conway, D.J. (2003), Evidence for diversifying selection on erythrocyte-binding antigens of *Plasmodium falciparum* and *P. vivax*, *Genetics*, **163**(4), 1327–1336.

Bergmann, S., Ihmels, J., and Barkai, N. (2003), Iterative signature algorithm for the analysis of large-scale gene expression data, *Phys. Rev. E Stat. Nonlin. Soft Matter. Phys.*, **67**(3), 031902-1-18.

Bozdech, Z., Llinas, M., Pulliam, B.L., Wong, E.D., Zhu, J.C., and DeRisi, J.L. (2003), The transcriptome of the intraerythrocytic developmental cycle of *Plasmodium falciparum*, *PLoS Biol.*, **1**(1), 85–100.

Conlon, E.M., Liu, X.S., Lieb, J.D., and Liu, J.S. (2003), Integrating regulatory motif discovery and genome-wide expression analysis, *Proc. Natl. Acad. Sci. USA*, **100**(6), 3339–3344.

Coulson, R.M., Hall, N., and Ouzounis, C.A. (2004), Comparative genomics of transcriptional control in the human malaria parasite *Plasmodium falciparum*, *Genome Res.*, **14**(8), 1548–1554.

Crooks, G.E., Hon, G., Chandonia, J.M., and Brenner, S.E. (2004), WebLogo: A sequence logo generator, *Genome Res.*, **14**(6), 1188–1190.

Eisen, M.B., Spellman, P.T., Brown, P.O., and Botstein, D. (1998), Cluster analysis and display of genome-wide expression patterns, *Proc. Natl. Acad. Sci. USA*, **95**(25), 14863–14868.

Gardner, M.J., Hall, N., Fung, E., White, O., Berriman, M., Hyman, R.W., Carlton, J.M., Pain, A., Nelson, K.E., Bowman, S., Paulsen, I.T., James, K., Eisen, J.A., Rutherford, K., Salzberg, S.L., Craig, A., Kyes, S., Chan, M.S., Nene, V., Shallom, S.J., Suh, B., Peterson, J., Angiuoli, S., Pertea, M., Allen, J., Selengut, J., Haft, D., Mather, M.W., Vaidya, A.B., Martin, D.M., Fairlamb, A.H., Fraunholz, M.J., Roos, D.S., Ralph, S.A., McFadden, G.I., Cummings, L.M., Subramanian, G.M., Mungall, C., Venter, J.C., Carucci, D.J., Hoffman, S.L., Newbold, C., Davis, R.W., Fraser, C.M., and Barrell, B. (2002), The genome sequence of the human malaria parasite *Plasmodium falciparum*, *Nature*, **419**(6906), 498–511.

Hall, N., Karras, M., Raine, J.D., Carlton, J.M., Kooij, T.W., Berriman, M., Florens, L., Janssen, C.S., Pain, A., Christophides, G.K., James, K., Rutherford, K., Harris, B., Harris, D., Churcher, C., Quail, M.A., Ormond, D., Doggett, J., Trueman, H.E., Mendoza, J., Bidwell, S.L., Rajandream, M.A., Carucci, D.J., Yates III, J.R., Kafatos, F.C., Janse, C.J., Barrell, B., Turner, C.M., Waters, A.P., and Sinden, R.E. (2005), A comprehensive survey of the Plasmodium life cycle by genomic, transcriptomic, and proteomic analyses, *Science*, **307**(5706), 82–86.

Han, J. and Kamber, M. (2001), *Data Mining: Concepts and Techniques*, Morgan Kaufmann Publishers.

Horrocks, P., Dechering, K., and Lanzer, M. (1998), Control of gene expression in *Plasmodium falciparum*, *Mol. Biochem. Parasitol.*, **95**(2), 171–181.

Hughes, J.D., Estep, P.W., Tavazoie, S., and Church, G.M. (2000), Computational identification of cis-regulatory elements associated with groups of functionally related genes in *Saccharomyces cerevisiae*, *J. Mol. Biol.*, **296**(5), 1205–1214.

Ihmels, J., Friedlander, G., Bergmann, S., Sarig, O., Ziv, Y., and Barkai, N. (2002), Revealing modular organization in the yeast transcriptional network, *Nat. Genet.*, **31**(4), 370–377.

Ihmels, J., Bergmann, S., and Barkai, N. (2004), Defining transcription modules using large-scale gene expression data, *Bioinformatics*, **20**(13), 1993–2003.

Liu, X.S., Brutlag, D.L., and Liu, J.S. (2002), An algorithm for finding protein-DNA binding sites with applications to chromatin-immunoprecipitation microarray experiments, *Nat. Biotechnol.*, **20**(8), 835–839.

Militello, K.T., Dodge, M., Bethke, L., and Wirth, D.F. (2004), Identification of regulatory elements in the *Plasmodium falciparum* genome, *Mol. Biochem. Parasitol.*, **134**(1), 75–88.

Rosenthal, P.J. (2004), Cysteine proteases of malaria parasites, *Int. J. Parasitol.*, **34**(13–14), 1489–1499.

Roth, F.P., Hughes, J.D., Estep, P.W., and Church, G.M. (1998), Finding DNA regulatory motifs within unaligned noncoding sequences clustered by whole-genome mRNA quantitation, *Nat. Biotechnol.*, **16**(10), 939–945.

Sturn, A., Quackenbush, J., and Trajanoski, Z. (2002), Genesis: Cluster analysis of microarray data, *Bioinformatics*, **18**(1), 207–208.

Tamayo, P., Slonim, D., Mesirov, J., Zhu, Q., Kitareewan, S., Dmitrovsky, E., Lander, E.S., and Golub, T.R. (1999), Interpreting patterns of gene expression with self-organizing maps: Methods and application to hematopoietic differentiation, *Proc. Natl. Acad. Sci. USA*, **96**(6), 2907–2912.

Thompson, J.K., Triglia, T., Reed, M.B., and Cowman, A.F. (2001), A novel ligand from *Plasmodium falciparum* that binds to a sialic-acid-containing receptor on the surface of human erythrocytes, *Mol. Microbiol.*, **41**(1), 47–58.

Troyanskaya, O., Cantor, M., Sherlock, G., Brown, P., Hastie, T., Tibshirani, R., Botstein, D., and Altman, R.B. (2001), Missing value estimation methods for DNA microarrays, *Bioinformatics*, **17**(6), 520–525.

van Helden, J. (2003), Regulatory sequence analysis tools, *Nucleic Acids Res.*, **31**(13), 3593–3596.

Voss, T.S., Kaestli, M., Vogel, D., Bopp, S., and Beck, H.P. (2003), Identification of nuclear proteins that interact differentially with *Plasmodium falciparum var* gene promoters, *Mol. Microbiol.*, **48**(6), 1593–1607.

Voss, T.S., Thompson, J.K., Waterkeyn, J., Felger, I., Weiss, N., Cowman, A.F., and Beck, H.P. (2000), Genomic distribution and functional characterisation of two distinct and conserved *Plasmodium falciparum var* gene 5′ flanking sequences, *Mol. Biochem. Parasitol.*, **107**(1), 103–115.

Chapter 10

Linking Gene Expression Patterns and Transcriptional Regulation in *Plasmodium falciparum*

Aidan J. Peterson, Andrew V. Kossenkov and Michael F. Ochs

Fox Chase Cancer Center, Philadelphia, PA 19111, USA

Abstract   Elucidation of the genome sequence of *P. falciparum*, the primary causative agent of human malaria, has opened new avenues for exploring the biology of this important microorganism. The CAMDA 2004 dataset offers a detailed view of mRNA transcript levels during the intra-erythrocyte stage of the parasite life cycle. Using Bayesian Decomposition to model expression patterns in the time series data, we examined the results over a range of potential solutions with the goal of choosing a number of patterns that modeled the experimental data faithfully, with genes in the patterns linked to biological processes. When the data was modeled with seven or eight patterns, each pattern represented a smooth temporal expression pattern whose contributing genes were enriched for Gene Ontology (GO) terms. As control of gene expression has not been elucidated in *P. falciparum*, we must work backwards from microarray profiles that represent the output of the transcriptional program. We use the upstream genomic sequences of genes linked by Bayesian Decomposition to uncover elements related to stage-specific transcriptional control. Sequence analysis revealed many motifs enriched in the temporal gene sets, but simulations revealed that enriched motifs are readily found in random sets of *P. falciparum* promoters. We therefore employed an enrichment factor ranking to focus on those motifs correlated with temporal phases. This analysis reveals a handful of candidate binding sites for transcription factors driving the *P. falciparum* erythrocytic cycle.

Keywords: microarray, gene expression, transcriptional regulation, Bayesian methods

1.   INTRODUCTION

Malaria is caused by *Plasmodium* parasites that infect and destroy several human cell types during their life cycle. The global effort to reduce the impact of malaria is intimately tied to the study of the complex biology of this protozoan parasite. The genome sequence for *P. falciparum*, the species responsible for the majority of malaria cases, was released in 2002 (Gardner et al., 2002), permitting genome-scale efforts to catalog the proteome and transcriptome of *Plasmodium* (Florens et al., 2002; Bozdech et al., 2003; Le Roch et al., 2003). One fundamental area of *Plasmodium* biology that remains poorly characterized is the control of gene expression, including regulation of gene transcription. The CAMDA 2004 dataset provides a high quality representation of the transcriptional behavior for most of the known and predicted genes of *P. falciparum* during the intra-erythrocyte development cycle (IDC). We wish to discover patterns in the expression data that are likely to reflect the action of biochemical mechanisms that generate biological co-regulation by affecting the expression of many genes. Such groups of co-regulated genes can then be used to explore regulatory features of the linked genes as an approach to refine understanding of the transcriptional control logic of *Plasmodium*. A better understanding of the biological processes that *Plasmodium* uses during the IDC should lead to improved therapeutics and eventual amelioration of this devastating disease.

The microarray data of Bozdech et al. (2003) and Le Roch et al. (2003) reveal robust variation in transcript levels through the IDC. Microarray results provide the transcriptional program *output*, yet we know very little about the *inputs* directing transcription in this organism. At the genomic level, the transcriptional profiles of the vast majority of *Plasmodium* genes do not show a discernable relationship to chromosome position, which suggests that regulatory mechanisms act on individual genes. Several promoters have been studied to determine which regions in the sequence control expression of a reporter transcript, and upstream control regions required for gene expression have been identified (Osta et al., 2002; Voss et al., 2003; Militello et al., 2004). In addition, several of these studies have detected DNA-binding activities from *Plasmodium* nuclear extracts (Osta et al., 2002; Voss et al., 2003). The proteins providing the basal transcription machinery are present in the genome, as are chromatin components and proteins containing motifs commonly associated with chromatin regulation. Sequence analysis of the *P. falciparum* genome, however, has exposed a conspicuous paucity of recognizable transcription factors (Aravind et al., 2003; Coulson et al., 2004).

The most conservative model drawn from the available data is that *P. falciparum* uses a set of DNA-binding proteins to control gene expression, but that these factors have not yet been identified experimentally or observed in the genome sequence. Indeed, a recent report describes a *P. falciparum* transcription factor identified based on distant protein similarity (Boschet et al., 2004). It has been suggested that *Plasmodium* relies heavily on post-transcriptional mechanisms to control protein expression. The most direct support for this notion is the discrepancy between the stage-specific qualities of the proteome (Florens et al., 2002) and the expression of the majority of genes in a single life cycle phase (Bozdech et al., 2003). Post-transcriptional control would have to operate *in addition to* the robust transcriptional control revealed by microarray studies, and should not interfere with our goal of identifying regulatory DNA sequences associated with co-expressed genes that may function in transcriptional control.

2. METHODS

2.1. Bayesian Decomposition

Analysis of the microarray data was done with Bayesian Decomposition (BD) (Moloshok et al., 2002) in order to identify key regulatory time points within the data. BD is a data analysis technique that decomposes a data matrix into an amplitude matrix and a pattern matrix such that the product of the two matrices models the input data. BD employs a computationally-intensive approach that samples many possible configurations of the amplitude and pattern matrices, and migrates towards solutions that best model the data. For gene expression data, each row of the pattern matrix represents a pattern of gene expression across the experimental conditions, and each row of the amplitude matrix indicates the loading applied to each pattern such that the weighted sum of the patterns approximates the observed expression for that gene. The number of patterns is defined by the user for each run of the algorithm. BD allows the identification of overlapping patterns within the data (here, overlapping times of expression) linked to specific genes. This permits the algorithm to both identify groups of genes that initiate expression, while other genes continue ongoing expression, and to identify genes that are regulated at multiple points during the parasitic life cycle. This is critically important for promoter analysis, since the limits of the genetic alphabet (G, A, T, C) make identification of DNA binding motifs difficult. The inclusion of promoter regions not truly involved in transcription factor binding quickly leads to loss of signal for identification of promoter elements and identification of additional false positives, a well known problem with promoter analysis (Bulyk, 2003).

BD was applied to the Overview data set comprising measurements of mRNA levels for 3719 oligos at 46 separate time points varying from 1 to 48 hours post infection (hpi). Expression levels were provided as ratios between the mRNA level at the time point and a pooled reference sample composed of a mixture of mRNA from all time points. Time points at 23 and 29 hours were removed by Bozdech et al. (2003) due to quality control problems. BD was run positing 3 to 12 patterns to permit analysis across a range of solutions. Duplicate modeling runs were performed using different random seeds in the Markov chain sampler. Results from the two independent runs were virtually identical except for the eight pattern condition; additional modeling runs with eight patterns were performed, and the two stable solutions were recovered multiple times (not shown). Computation time for a BD run depends on the size of the data matrix and the number of solution patterns; the range of analysis run times in this study was approximately 30 minutes to 16 hours. We estimated the noise at 20% of signal (i.e., a multiplicative noise). This estimate is higher than the variations observed in the limited number of replicates performed by Bozdech et al. (2003), but within a range where BD results are not sensitive to the value of the noise estimate (Moloshok et al., 2002). Missing values were assigned a ratio of 1 with an uncertainty of 100, allowing the algorithm to essentially ignore their contribution during modeling.

2.2.   Pattern Visualization

Output files from BD analyses were imported into the ClutrFree (Bidaut and Ochs, 2004) program to visualize the expression pattern elements and to analyze the relationships between the patterns. ClutrFree was also used to export the membership matrices describing the weights of each pattern assigned to each oligo element to reconstitute its overall expression profile.

The published Gene Ontology (GO) term annotations for the *P. falciparum* genome (Ashburner et al., 2000; Gardner et al., 2002) were combined with the member lists for each pattern determined by BD. For each set of patterns, the ClutrFree program displays the enrichment of GO terms in each pattern and an associated *p*-value based on a hypergeometric distribution (Bidaut and Ochs, 2004). A simple metric was devised to track the GO term enrichment strength for each pattern: Process and Function terms enriched with *p*-values less than 0.01 were assigned two points, and those with *p*-values between 0.01 and 0.05 were assigned one point, and the sum of the points was defined as the enrichment score for the pattern. The average score per pattern was used to compare enrichment strength for a range of fitted patterns. The metric used here is not intended as a thorough analysis of GO enrichment, since it does not use particularly stringent *p*-value cutoffs and does not consider that many genes are annotated at several GO levels. It is a valid tool to compare different analyses of this data, however, since annotation bias and false positives should apply similarly to each condition.

2.3.   Identification of Regulatory Sequences

To discover potential regulatory DNA sequences related to temporal gene expression, we used the pattern information from BD to place genes into co-regulation groups. A list of oligos with strong membership in each of eight patterns was converted to a gene list, and the genomic sequence near the annotated gene sequence was extracted from PlasmoDB data files (Bahl et al., 2003). In some cases, two or more oligos map to the same gene. For these genes, the

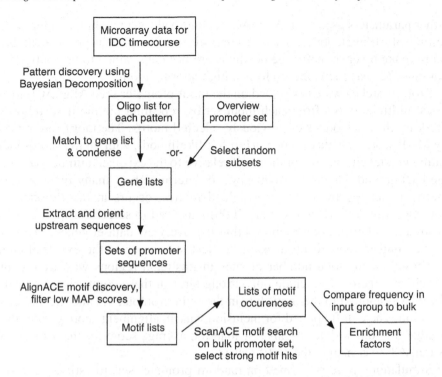

Figure 1. Diagram of the workflow to define sequence motifs associated with expression patterns in *P. falciparum* IDC gene expression. Gene lists were generated by association with BD expression patterns, or randomly selected from the bulk set of promoter region. Details for each step are found in the text.

average membership percentage was used unless the membership percentages differed by more than 20%, in which case the higher values were used. The ATG translational start codon documented on PlasmoDB was used as a reference point since very few transcription start sites have been mapped. A larger set of upstream DNA sequences was generated from a list of all genes represented in the Overview dataset. After excluding rRNA, tRNA and organelle-encoded genes, this control set of upstream regions contains 2683 sequences. The upstream sequences were oriented in the direction of gene transcription prior to sequence analysis. Connections between the pattern detection and motif detection steps of our strategy are outlined in Figure 1.

The AlignACE program was used to analyze promoter sets to find enriched sequences. AlignACE is an implementation of a Gibbs sampling approach that finds motifs that are found more often than expected based on nucleotide frequencies (Roth et al., 1998; Hughes et al., 2000). For the 2 kb upstream sequences, GC content was modeled as 0.15 to match the actual GC content of the 2683 two kilobase promoter regions represented in the overview dataset.

Other parameters used were AlignACE default values, including column width setting of 10 nucleotides. The ten nucleotide positions that provide the best score make up each motif; the positions are not constrained to be contiguous, but must be near each other to form a high-scoring motif.

Motif searches were performed on the input promoter sets. AlignACE identifies multiple motifs from each input set by repeating the motif search after masking the nucleotides of previously selected motifs. The motifs are ranked by MAP score (the maximum *a priori* log-likelihood), with higher scores indicating greater enrichment of the motif relative to the expected number based on the background nucleotide frequency. To determine how many times a given motif, as defined by a pattern weighted matrix, occurs in the *Plasmodium* genome, ScanACE (Hughes et al., 2000) was used to search for motifs in the upstream sequence library representing the overview dataset genes. The ratio of the number of motif instances in the test subset versus the expected number based on the total number of sites in upstream regions was used as the enrichment factor. To include only strong sites in these comparisons, a score cutoff was applied to the motif hits in the bulk promoter set and the subset in question. The cutoff applied for inclusion was an alignment score greater than a value 0.5 standard deviations less than the average score for the set used by AlignACE to construct the motif matrix.

Simulations were performed on random promoter sets to estimate the significance of the enrichment factors determined for the promoter sets linked by co-expression. For several different group sizes, random sets were assembled from the bulk promoter sequences, disallowing a sequence to occur more than once in a given random group. These groups were then analyzed by AlignACE to identify motifs, and the enrichment factors were determined following ScanACE analysis of the bulk promoter set. WebLOGO representations of the top motifs were generated from the Web server (Crooks et al., 2004).

3.   RESULTS

3.1.   Temporal Expression Peaks

BD was performed to find the prominent component patterns in the expression profiles of the oligo elements in the overview dataset of Bozdech et al. (2003). Although the appropriate number of patterns to fit for a particular data set can only be judged retrospectively (see below), we initially chose a range of 3 to 12 patterns to carry out the analysis. The lower number is based on the expectation that there is at least one regulatory phase for each major morphological stage of *Plasmodium* during the blood cell cycle, and the upper number is arbitrary but comfortably above the 6 patterns used in the BD analysis of the yeast cell cycle (Moloshok et al., 2002). Figure 2 shows a pattern tree representation from ClutrFree that diagrams the correlations across the patterns. In this

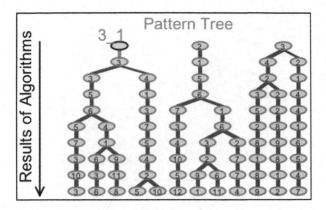

Figure 2. Pattern tree relating expression patterns found by BD for the IDC gene expression cycle. Shown here for 3 to 12 patterns, the ClutrFree tree links the most related gene expression patterns, providing a way to visualize the stability and splitting of the patterns as the number of fitted patterns is increased. Moving vertically down a chain, a single connection indicates two patterns are similar in two modeling runs. A branch indicates that two gene expression patterns are most similar to a single pattern among those determined for a smaller number of patterns. The branch points serve to highlight the main difference in the result when an additional expression pattern is allowed during the modeling run.

depiction, linked spots represent the gene expression patterns that are most similar to each other between different BD runs. Branch points highlight significant changes between modeling runs. The pattern tree representation thus serves as a navigation tool that allows the examination of the shapes of individual gene expression patterns as well as the differences between solutions with different number of patterns. This analysis permits a qualitative overview of how BD mathematically models the *Plasmodium* gene expression data.

The majority of the individual expression patterns appear as unimodal curves distributed along the time course, despite the fact that no smoothing function was used by the algorithm (examples shown in Figure 3). Patterns with peaks closest to the first and last time points have a single peak if the pattern is considered to wrap to the next cycle of the IDC. Visual analysis of the patterns revealed that as the number of patterns fit by the algorithm increased, "parent" patterns split into two patterns, each with temporal peaks offset to either side of the peak of the parent pattern. An example is shown in Figure 3, comparing one of six patterns to the two most related of seven patterns. As the number of patterns increased beyond eight, they began to exhibit features that are unlikely to reflect true temporal gene expression patterns. For example, one of the nine patterns has a broad peak composed of amplitudes that are erratic from hour to hour. Biological patterns are expected to be smooth because the cell cycle synchrony of *Plasmodium* cells in the raw biological material is not exact, so the erratic peaks almost certainly reflect overfitting by the algorithm.

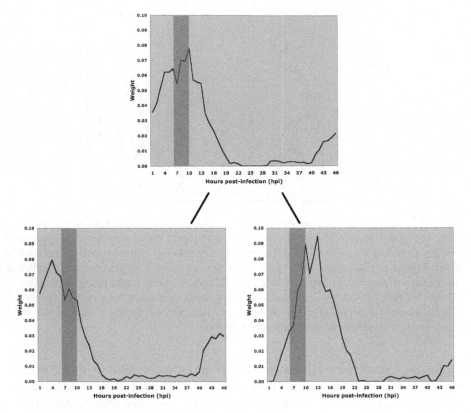

Figure 3. Representative patterns showing relationship between "parent" and "children" temporal patterns. In this example, an expression pattern from the 6 pattern set (top graph) and two patterns from the 7 pattern set (middle and bottom graphs) are shown. The pattern numbers are arbitrary in the sense that the patterns are found de novo during each BD run, but they correspond to patterns 6 of 6, 7 of 7 and 1 of 7 on the pattern tree in Figure 2, where they occupy a branch point in the tree. The peak centered near 8 hpi splits into two peaks near 4 and 12 hpi when an additional pattern is fit by BD. The shaded bar is at the same position of each plot to highlight peak shifts.

Our conclusion is that fewer than nine patterns should be used to fit the data since some of the additional patterns are less likely to represent true biology.

For three through seven patterns, different simulation runs yielded virtually identical results, indicating that the solutions are robust (i.e. the same stable solution is reached starting from different random starting points in the model space). The stable positions and shapes of the temporal patterns indicate that the individual expression patterns are not uniformly distributed across the IDC time course. With eight patterns, however, one solution invoked a pattern resembling the "noisy" pattern from the 9 pattern set, but another solution split a temporal peak into smooth daughter peaks, as was observed for pattern number increases in the lower range. This indicates that the sampling algorithm is able to identify two mathematically acceptable solutions, as seen in some other contexts (Ochs et al., 1999). In summary, the fit patterns are stable over different runs out to seven patterns, and above this number, alternate solutions become possible and thus are found in different fitting runs. This observation suggests that 7 or 8 patterns is the appropriate range to take advantage of the robust temporal patterns in the data without including redundant or artifactual expression patterns.

3.2.   Gene Ontology Enrichment

In addition to the visual inspection of pattern features described above, we considered how well the various patterns clustered genes with related gene functions. Approximately 40% of *P. falciparum* genes have GO annotations that indicate their biological roles (Gardner et al., 2002). We examined the GO terms that are enriched (i.e. found more often than expected based on frequency of the GO term in the entire genome) for the expression patterns modeled by BD to determine if they represent meaningful biological themes. For example, many of the enriched terms derived from the solution with seven solution patterns make sense in terms of the invasion, replication, and maturation cycle that occurs inside the erythrocytes. Genes related to metabolism, energy generation, and protein synthesis dominate immediately after invasion, followed by DNA replication prior to cell division, and finally the schizonts express transporters, kinases, and surface molecules in preparation for the next round of invasion. Enrichment of other features, such as RNA binding and mRNA processing terms, is not easily explained, but suggests broad areas of study that may provide insight into *Plasmodium* replication during the IDC.

GO term enrichment has been noted for temporal gene expression groups in the published CAMDA data set (Bozdech et al., 2003) and in clustered gene expression profiles detected with a short-oligo array set (Le Roch et al., 2003). Collectively, these results support the notion that the control of gene expression in *P. falciparum* leads to frequent co-regulation of related genes. We determined the number of GO terms enriched above significance thresholds for each pattern in each set of pattern solutions. Many GO terms are enriched regardless of the number of patterns. In order to use this information to guide the selection of the number of expression patterns to fit, we calculated the average number of enriched terms per pattern, over the range of 3 to 12 patterns. A plot of this GO enrichment score versus number of expression patterns (Figure 4) shows that the enrichment per pattern peaks at 7 to 8 patterns, and becomes erratic with higher pattern numbers. To the extent that GO term clustering represents meaningful co-regulation, this analysis suggests that 7 or 8 patterns present biologically motivated solutions. This consideration is especially important if the groups defined by clustering will be used to search for novel

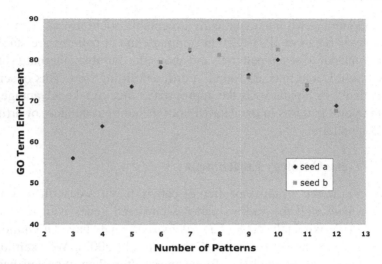

Figure 4. GO term enrichment as a function of the number of basic vectors in Bayesian Decomposition. The average GO enrichment per pattern increases as additional patterns are allowed from 3 to 8 patterns, indicating that the newly formed patterns better fit the biological grouping of the genes. At 9 patterns and above, the average enrichment declines, indicating that the additional finer patterns have less correspondence to biological groups. Seeds a and b indicate different random starting points for fitting the same data.

shared features such as regulatory DNA sequence elements in co-regulated genes. For promoter region analysis, we chose the 8 pattern BD solution that consisted entirely of smooth temporal peaks and had the highest average GO term enrichment, since this single solution best fulfilled both of the selection criteria.

3.3.   Pattern Membership Distributions

The eight expression patterns provide coverage for the entire 48 hour experimental time course of the IDC. This is expected since all points along the time course feature expression of many oligo elements, and the algorithm seeks to find patterns that can model all of the input data. In the phasogram presented in Figure 2 of Bozdech et al. (2003), the expression peaks appear to be smoothly distributed along the IDC time course. The expression motif peaks modeled by BD, however, are not evenly distributed (Table 1). Furthermore, the extent that each pattern is used in the BD amplitude matrix varies from 5% for the 47 hpi pattern to 24% for the 25 hpi pattern (Table 1). Our central hypothesis is that a limited set of transcriptional regulation phases drive the expression behavior. Since the eight expression patterns produced by BD are derived directly from the data, we chose these patterns as the source for gene groups to discover regulatory elements, rather than selecting arbitrary or evenly-spaced time windows within the IDC.

Table 1. Genes and motifs associated with the eight temporal expression patterns determined by Bayesian decomposition. The peaks are arranged chronologically in the IDC according to the approximate midpoints of the pattern peak. The total weight of a pattern represents the cumulative usage of that pattern to explain the gene expression profiles, with values reported as per cent of total. The number of genes strongly linked to each pattern is shown, using three different cutoffs for inclusion. The >60% group, where each gene tallied has more than 60% of its behavior explained by a pattern, was primarily used to search for promoter sequence motifs. Motifs represents the number of high-scoring motifs identified by AlignACE. Blank entries indicate that AlignACE analysis was not performed on a promoter list. A subset of these motifs passes the significance filters described in the text

Peak:	4 hpi	11 hpi	18 hpi	25 hpi	33 hpi	38 hpi	43 hpi	47 hpi
Total weight of pattern	0.10	0.16	0.10	0.24	0.17	0.11	0.08	0.05
Genes, >50%	31	97	58	460	328	128	85	10
Motifs		13						
Pass filters		0						
Genes, >60%	14	32	28	276	177	72	62	4
Motifs	4	7	9	50	29	12	15	4
Pass filters	0	0	2	2	1	3	3	0
Genes, >75%	8	9	12	84	41	30	25	3
Motifs				25				
Pass filters				1				

For each of the eight expression patterns, strong member genes were selected based on the percentage of gene expression behavior explained by each pattern. We considered different stringencies for defining the representative genes for each pattern. Table 1 shows how many genes have greater than 50, 60 or 75% behavior explained by one of the 8 patterns. For each cutoff, the number of surviving genes in each pattern varies widely. For example, at the 60% cutoff, two patterns have fewer than 20 strong member genes, whereas two of the patterns have more than 150 strong members. The statistical strength of the AlignACE algorithm is best utilized with a moderate number of input sequences, in the range of several dozen to several hundred (Hughes et al., 2000). In practice, computation time slows dramatically when hundreds of long input sequences are used. We therefore chose to perform the main analysis on the 60% membership cutoff.

3.4. Enriched Sequence Motifs

We used AlignACE to search for motifs in the sets of upstream sequences of strong member genes for each BD pattern. Table 1 shows the number of motifs with a MAP score greater than 10 for each pattern. A MAP score of 10 was chosen for the cutoff based on extensive tests performed in yeast (Hughes et al., 2000) where known biological sites have scores greater than 10. The number ranges from 4 motifs for the 4 hpi and 47 hpi patterns, to 50 motifs for the 25 hpi pattern. It is expected that there are many false positives motifs in this group, so an important phase of the analysis is applying filters and criteria to select the most reasonable set of motifs.

The top scoring motifs for each pattern were AT repeats and poly-A tracts, which is not surprising since these simple repeats are common in the *P. falciparum* genome. Since AlignACE models the background sequence as a simple A/T percentage (i.e. a zero order relationship between positions), these sequences score well, despite their obvious lack of specificity to a given input set of promoters. Additional motifs found by AlignACE showed greater sequence variety, and represent more plausible transcription factor binding sites. Another feature of the motifs identified by AlignACE is that some of the motifs retrieved from an input set are similar or nearly identical to each other. This situation can be addressed by clustering similar motifs to avoid redundancy in later analysis steps (Hughes et al., 2000). We found that a simple filter based on group specificity addressed both of these concerns since the motifs showing intra-group similarity tended to be the same motifs found in multiple groups.

The key variable in our analysis is variation in gene expression over time, so the primary goal of the sequence analysis is to identify motifs associated with the various temporal expression patterns. To determine the extent of enrichment of a motif for an input set, we compared the number of occurrences of that motif in the set to the expected number of occurrences. The expected number was determined by scanning the promoter regions for all genes in the overview data set, and calculating the enrichment factor as the observed number of motifs divided by the expected number for an input set of that size. To determine the number of motif instances, we included only strong motifs by applying a consistent motif score cutoff to the motif hits for the bulk promoter set and the promoter subset related to the motif.

3.5.   Significance of Motif Enrichment Factors

The majority of motifs identified from the pattern promoter sets have low enrichment factors, but a number of the motifs have high enrichment values. Of the 106 (unclustered) motifs under consideration, 79 have enrichment values less than 2, 10 have values between 2 and 4, and 16 have values greater than 4. To estimate the significance of these values, we simulated the analysis using sets of random promoter sequences derived from the overview gene set. The length of sequence in the original search groups varied because we used the BD pattern strength to select the groups. It is clear even from the eight groups tested that the range of enrichment values tends towards smaller values for the larger input sets. We therefore considered the size of the input sequence set as a key parameter in simulations of motif discovery and enrichment calculations.

Promoter sets of 14, 32, 50, and 72 sequences were chosen randomly from the bulk list of promoters represented in the overview dataset. For each set size, 24 groups were analyzed by AlignACE, and the motifs with MAP scores greater than 10 were carried through the ScanACE and enrichment factor calculations. Some of the motifs from random input sets had extreme enrichment factors (in the neighborhood of 50 fold enrichment over expected). Inspection of several of these motifs revealed that they were dominated by short direct repeats found in one or two promoters. These motifs have few hits in genomic upstream sequences, and very few hits at the gene level. To remove these apparently spurious motifs, we applied a cutoff requiring a motif to be found in more than 50 genes in the 2683 overview promoters. (50 genes would be a respectable number of targets, but even for known factor binding sites, the number of sequence occurrences is much larger than the number of functional regulatory sites.) The filtered enrichment values were pooled for each condition, and plotted versus the percentile rank. Figure 5 shows that the size of the input sequence set has a strong effect on the spectrum of enrichment scores. Enrichment scores were very high for the 14 and 32 sequence sets, such that the majority of enrichment values were greater than 4. The plots for input sets of size 50 and 72 are similar to each other, and have fewer motifs with very high enrichment scores. For example, only 15 per cent of the motifs are enriched more than four fold in the data from the promoter sets with 50 or more members.

We used the information from the random promoter set simulations to assign a percentile rank to the motifs mined from the temporal expression groups. The 47 hpi peak pattern was excluded for lack of member genes. For the 4 hpi and 11 hpi patterns represented by 14 and 32 genes, we used the respective percentile scales from the randomized sets. No motif from these groups ranked in the top quintile. The 18 hpi peak group had 28 promoter sequences in the selection group; the 32 promoter scale was used as an approximation but the concern remains than the limited number of input sequences undercuts the significance of these motifs. For the remaining groups, the sample size ranges from 62 to 276. We extrapolated from the behavior of the plots in Figure 5 that sets with greater than 50 input sequences should have similar profiles of enrichment values. While we did not simulate enough cases to verify this assumption, it fulfills the primary purpose of eliminating over-interpretation of motifs derived from small sample sets. The 50-promoter percentile scale was therefore applied to the remaining motifs. For each group, we selected the top motifs with enrichment factor percentile ranks in the top quintile as the best candidates for stage-specific regulatory elements. The final number of motifs

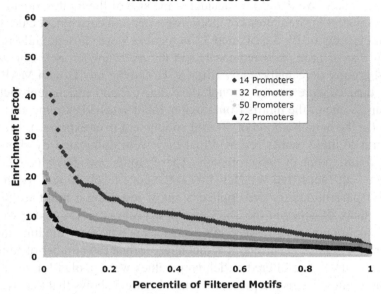

Figure 5. Enrichment factors from random groups of promoters. Motif searches were performed on random promoter sets of the indicated sizes. Results from multiple runs for each condition were pooled, after removing motifs with low MAP scores and those with very few occurrences. Enrichment factors are plotted against percentile to indicate how often high enrichment factors can be expected for input promoter sets of various sizes.

passing the series of filters is shown in Table 1, and illustrates that a minority of the "high-scoring" sequence motifs fulfill the requirements.

The properties of the top motifs are shown in Table 2. The average number of motif hits per gene in the input set is presented along with the enrichment factor for each motif. Note that even for the most significantly enriched motifs, the hits per gene is less than one. One explanation is that even within each expression group there is likely to be more than one active factor binding site, and thus the genes would not be expected to contain any single site. The motifs themselves are represented as sequence LOGOs (Crooks et al., 2004), which indicate the nucleotide preference for each position and the height of the letters reflects the weight of the position in the position weighted matrix. The width of the motifs varies from 11 to 23 nucleotides because AlignACE scores the ten strongest positions without requiring them to be contiguous. The LOGO displays include all positions in the motif, including those interspersed with the active positions. Some of the motifs have cores of strongly conserved residues, whereas some of the longer motifs comprise scattered weaker positions. For example, motif 43hpi-15 is dominated by a consensus sequence of GTGTGCA. Motif 38hpi-4 features a GTGCAC palindrome sequence consen-

Table 2. Top sequence motifs associated with expression patterns. The motifs are labeled by pattern and AlignACE motif number. Hits per gene indicates the average number of motifs found per gene in the input set of promoters. The rank attached to each enrichment factor indicates percentile rank relative to simulations with random promoters, as described in the text. Web logo depictions of each motif display the preferred nucleotides for each position. Letter height indicates the significance of the position; letter width varies with motif width

Motif name	Hits per gene	Enrichment(rank)	LOGO
18hpi-8	0.39	19(1)	Pattern 5 Motif 8
18hpi-9	0.57	17(1)	Pattern 5 Motif 9
25hpi-16	0.64	5.9(6)	pattern 4 motif 16 (of top 84)
25hpi-31	0.23	5.0(9)	Pattern 4 Motif 31
25hpi-29	0.26	4.6(11)	Pattern 4 Motif 29
33hpi-27	0.30	9.2(3)	Pattern 7 Motif 27
38hpi-4	0.50	5.9(7)	Pattern 1 Motif 4
38hpi-7	0.44	5.7(7)	Pattern 1 Motif 7
38hpi-12	0.21	4.8(10)	Pattern 1 Motif 12

Table 2. (Continued)

Motif name	Hits per gene	Enrichment(rank)	LOGO
43hpi-7	0.37	16(1)	
43hpi-9	0.24	5.7(4)	
43hpi-15	0.68	4.8(15)	

sus, which is notable because many known transcription factor binding sites are palindromes. At the other end of the spectrum, all three of the motifs from the 25 hpi pattern, 38hpi-7, and 43hpi-7 have several strong positions dispersed among weak positions. These motifs can not be excluded from consideration, but because they do not resemble features of known binding motifs, we consider them less likely to be genuine regulatory elements. The remaining top motifs have at least 5 or 6 positions with strong nucleotide preference, and therefore have enough specificity to potentially serve as regulatory binding sites for transcription factors.

4.   CONCLUSIONS

We used BD to discover gene expression pattern elements in the *P. falciparum* IDC transcriptome. We chose the BD approach over other clustering and pattern finding algorithms because we are primarily interested in finding gene expression features that may be linked to gene expression regulatory mechanisms. BD provides these expression features (i.e. solution patterns) as a primary output, whereas traditional clustering approaches output groups of genes, often requiring additional clusters to accommodate genes affected by multiple regulatory mechanisms. Reflecting the nature of the individual input elements, the most robust expression pattern elements in our results have a single prominent peak during the IDC. As with any pattern discovery method, choosing an appropriate number of patterns is not straightforward. We found that fitting seven or eight patterns produced smooth temporal patterns sufficient to reconstruct the data, and maximized the GO term enrichment of the patterns. These results are not meant to suggest that there are seven or eight regulatory events or transcription factors driving gene expression in the IDC, but that the identified temporal expression patterns are excellent places to explore co-regulation mechanisms. This rationale is inspired by the results of microarray analysis of the yeast cell cycle, where gene expression patterns are enriched for genes controlled by specific transcription factors that are required to drive the cell cycle (Spellman et al., 1998).

As one approach to explore potential transcriptional control programs in *P. falciparum*, we searched for sequence motifs that correspond to the various temporal expression patterns. Using the AlignACE program to detect overrepresented sequences, we were able to find many sequence motifs from the promoters of genes in each pattern. Motif discovery approaches in other organisms historically detect many false positive motifs. We addressed this concern for *P. falciparum* by performing a series of sequence motif mining simulations on random promoter sets. Over-represented motifs were readily detected from any collection of input sequences, indicating that merely finding a sequence motif is not meaningful. To highlight motifs that are specifically enriched in certain temporal expression patterns, we determined for each motif an enrichment factor comparing the number of motif occurrences in the group of interest to the number of occurrences expected based on the frequency in all promoter regions. Enrichment factor analysis of the motifs from random promoter sets led us to exclude motifs discovered from small input sets, as well as motifs that were very rare in all genomic promoter sequences. The handful of motifs that pass all these criteria represent our candidate list of binding sites for factors driving temporal gene expression. This approach is not designed to detect common or core promoter elements, since they are not expected to be enriched in specific expression profiles. A search for common elements associated with promoters would require comparing sequence from different positions with respect to genes.

There are a number of factors that confound the prediction of transcription factor binding sites. In this study we seek over-represented sequence motifs found in gene expression patterns because we expect transcription factor binding motifs to have this property. It is important to note that the motifs identified by sequence analysis alone are not constrained to match the sequence features recognized by a binding factor, and should therefore be expected to be loose, overlapping representations. Computational predictions of binding sites can be refined by considering clustering of motifs, and evolutionary conservation of homologous promoter sequences (Bulyk, 2003). These approaches should prove useful for sequence analysis in *P. falciparum* since a handful of *Plasmodium* genomes are being sequenced (Carlton et al., 2002; Hall et al., 2005). Another concern, clearly evident from the random promoter set simulations, is that detected motifs may be an artifact of the particular input sequences used for motif discovery. We used a simple analysis to gauge whether our top-scoring motifs were associated with the temporal patterns from which they were derived. For each of our top ten motifs, we found that genes linked to the pattern from which the motif was identified were more likely to contain the motif than a randomly chosen gene (not shown). The genes used as the original input for motif discovery were excluded since they were already known to be enriched for the motif. This trend is consistent with the hypothesis that there are sequence features that are enriched in groups of temporally co-expressed genes.

Validation of the motif sequences will require biological experiments to determine if they are required for transcriptional regulation, and to identify the transcription factors that bind the sites. Gene expression reporter assays that compare the activity of sequences with intact and disrupted motifs can be used to ascertain which motifs are functional. Nuclear extracts of *P. falciparum* proteins can be used to detect binding to various sequences. These approaches remain laborious, but it would be feasible to apply them to directly study the handful of candidate motifs associated with specific gene expression profiles in the IDC presented here. Motifs identified in this manner would contribute to our nascent understanding of transcription in *Plasmodium*. Additionally, positive results for some of the motifs would demonstrate the usefulness of motif prediction from sequence analysis, which will be particularly useful for organisms where genomic data is available but molecular biological studies are impractical.

ACKNOWLEDGEMENTS

This work was supported by NIH (NCI CA06927 and NCI P30CA06973), the Pennsylvania Department of Health, and the Pew Foundation. We thank Ghislain Bidaut for providing an updated version of the ClutrFree software, and Thomas Moloshok and Sinoula Apostolou for assistance and discussion.

REFERENCES

Aravind, L., Iyer, L.M., Wellems, T.E., and Miller, L.H. (2003), Plasmodium biology: Genomic gleanings, *Cell*, **115**, 771–785.

Ashburner, M., Ball, C.A., Blake, J.A., Botstein, D., Butler, H., Cherry, J.M., Davis, A.P., Dolinski, K., Dwight, S.S., Eppig, J.T., Harris, M.A., Hill, D.P., Issel-Tarver, L., Kasarskis, A., Lewis, S., Matese, J.C., Richardson, J.E., Ringwald, M., Rubin, G.M., and Sherlock, G. (2000), Gene ontology: Tool for the unification of biology. The Gene Ontology Consortium, *Nat. Genet.*, **25**, 25–29.

Bahl, A., Brunk, B., Crabtree, J., Fraunholz, M.J., Gajria, B., Grant, G.R., Ginsburg, H., Gupta, D., Kissinger, J.C., Labo, P., Li, L., Mailman, M.D., Milgram, A.J., Pearson, D.S., Roos, D.S., Schug, J., Stoeckert Jr., C.J., and Whetzel, P. (2003), PlasmoDB: The *Plasmodium* genome resource. A database integrating experimental and computational data, *Nucleic Acids Res.*, **31**, 212–215.

Bidaut, G. and Ochs, M.F. (2004), ClutrFree: Cluster tree visualization and interpretation, *Bioinformatics*, **20**, 2869–2871.

Boschet, C., Gissot, M., Briquet, S., Hamid, Z., Claudel-Renard, C., and Vaquero, C. (2004), Characterization of PfMyb1 transcription factor during erythrocytic development of 3D7 and F12 *Plasmodium falciparum* clones, *Mol. Biochem. Parasitol.*, **138**, 159–163.

Bozdech, Z., Llinas, M., Pulliam, B.L., Wong, E.D., Zhu, J., and DeRisi, J.L. (2003), The transcriptome of the intraerythrocytic developmental cycle of *Plasmodium falciparum*, *PLoS Biol.*, **1**, E5.

Bulyk, M.L. (2003), Computational prediction of transcription-factor binding site locations, *Genome Biol.*, **5**, 201.

Carlton, J.M., et al. (2002), Genome sequence and comparative analysis of the model rodent malaria parasite *Plasmodium yoelii yoelii*, *Nature*, **419**, 512–519.

Coulson, R.M., Hall, N., and Ouzounis, C.A. (2004), Comparative genomics of transcriptional control in the human malaria parasite *Plasmodium falciparum*, *Genome Res.*, **14**, 1548–1554.

Crooks, G.E., Hon, G., Chandonia, J.M., and Brenner, S.E. (2004), WebLogo: A sequence logo generator, *Genome Res.*, **14**, 1188–1190.

Florens, L., Washburn, M.P., Raine, J.D., Anthony, R.M., Grainger, M., Haynes, J.D., Moch, J.K., Muster, N., Sacci, J.B., Tabb, D.L., Witney, A.A., Wolters, D., Wu, Y., Gardner, M.J., Holder, A.A., Sinden, R.E., Yates, J.R., and Carucci, D.J. (2002), A proteomic view of the *Plasmodium falciparum* life cycle, *Nature*, **419**, 520–526.

Gardner, M.J., et al. (2002), Genome sequence of the human malaria parasite *Plasmodium falciparum*, *Nature*, **419**, 498–511.

Hall, N., et al. (2005), A comprehensive survey of the *Plasmodium* life cycle by genomic, transcriptomic, and proteomic analyses, *Science*, **307**, 82–86.

Hughes, J.D., Estep, P.W., Tavazoie, S., and Church, G.M. (2000), Computational identification of cis-regulatory elements associated with groups of functionally related genes in *Saccharomyces cerevisiae*, *J. Mol. Biol.*, **296**, 1205–1214.

Le Roch, K.G., Zhou, Y., Blair, P.L., Grainger, M., Moch, J.K., Haynes, J.D., De La Vega, P., Holder, A.A., Batalov, S., Carucci, D.J., and Winzeler, E.A. (2003), Discovery of gene function by expression profiling of the malaria parasite life cycle, *Science*, **301**, 1503–1508.

Militello, K.T., Dodge, M., Bethke, L., and Wirth, D.F. (2004), Identification of regulatory elements in the *Plasmodium falciparum* genome, *Mol. Biochem. Parasitol.*, **134**, 75–88.

Moloshok, T.D., Klevecz, R.R., Grant, J.D., Manion, F.J., Speier, W.Ft., and Ochs, M.F. (2002), Application of Bayesian decomposition for analysing microarray data, *Bioinformatics*, **18**, 566–575.

Ochs, M.F., Stoyanova, R.S., Arias-Mendoza, F., and Brown, T.R. (1999), A new method for spectral decomposition using a bilinear Bayesian approach, *J. Magn. Reson.*, **137**, 161–176.

Osta, M., Gannoun-Zaki, L., Bonnefoy, S., Roy, C., and Vial, H.J. (2002), A 24 bp cis-acting element essential for the transcriptional activity of *Plasmodium falciparum* CDP-diacylglycerol synthase gene promoter, *Mol. Biochem. Parasitol.*, **121**, 87–98.

Roth, F.P., Hughes, J.D., Estep, P.W., and Church, G.M. (1998), Finding DNA regulatory motifs within unaligned noncoding sequences clustered by whole-genome mRNA quantitation, *Nat. Biotechnol.*, **16**, 939–945.

Spellman, P.T., Sherlock, G., Zhang, M.Q., Iyer, V.R., Anders, K., Eisen, M.B., Brown, P.O., Botstein, D., and Futcher, B. (1998), Comprehensive identification of cell cycle-regulated genes of the yeast *Saccharomyces cerevisiae* by microarray hybridization, *Mol. Biol. Cell*, **9**, 3273–3297.

Voss, T.S., Kaestli, M., Vogel, D., Bopp, S., and Beck, H.P. (2003), Identification of nuclear proteins that interact differentially with *Plasmodium falciparum var* gene promoters, *Mol. Microbiol.*, **48**, 1593–1607.

Chapter 11

Chromosomal Spatial Correlation of Gene Expression in *Plasmodium falciparum*

J.B. Christian[a], C. Shaw[c], J. Noyola-Martinez[a], M.C. Gustin[b], D.W. Scott[a]
and R. Guerra[a,*]

[a]*Department of Statistics, Rice University, Houston, TX 77005, USA*
[b]*Department of Biochemistry and Cell Biology, Rice University, Houston, TX 77005, USA*
[c]*Department of Molecular and Human Genetics, Baylor College of Medicine,
Houston, TX 77030, USA*

Abstract   Malaria is responsible for half a billion infections and two million deaths each year. Understanding the biology of *Plasmodium falciparum* is critical if effective vaccines are to be developed to fight against this aggressive parasite. New information about the regulatory mechanisms of *P. falciparum* promotes the elucidation of the fundamental metabolic and transcriptional pathways which we must understand to design vaccines and better treatments. Of particular importance is the intraerythrocytic development cycle (IDC), the part of the *P. falciparum* life cycle spent in the blood stream of host mammals and that is responsible for the physical symptoms of malaria. The goal of this investigation is to examine spatially dependent co-regulation of gene expression over the 48-hour IDC. Correlation between gene expression and gene location over a few genes demonstrates evidence of co-regulated genes or operons, while correlation over many genes may provide evidence for some other transcriptional regulation mechanism such as chromatin remodeling or enhancers. We develop and apply a visualization and statistical testing methodology to examine expression–location correlation in a time-course microarray study of the IDC transcriptome. Contrary to the current paucity of evidence, our findings show evidence for spatial correlation. The biological implications of detected blocks of moderate but consistent spatial correlation provide novel insights into the transcriptome of *P. falciparum*.

Keywords: co-regulation, spatial correlation, DNA sequence data, microarray data, integration of data sources, visualization, permutation tests

---

*Corresponding author.

1.   INTRODUCTION

Understanding the regulatory mechanisms in *P. falciparum* helps identify new targets for both preventing or stopping malaria infections. The study of transcriptional regulation is paramount to achieving these goals, and there are many interesting transcriptional phenomena in *Plasmodium*. With secondary, tertiary and quaternary levels of structure in the DNA, there is much speculation about the randomness of the ordering of genes. Operons, chromatin remodeling and enhancers can affect gene transcription over short, mid and long distances, respectively, along the chromosome. In addition, protozoa such as *Plasmodium* are capable of regulating gene expression by altering their chromosome structure. For example, expression of the *var* cell surface protein of *Plasmodium* is regulated by a silencing mechanism [7]. In other eukaryotes, gene silencing and related epigenetic phenomena are typically mediated by covalent modification of histones that can spread along chromosomes, altering the accessibility of genes to the transcription apparatus [14]. Whether this type of regulation extends beyond the *var* genes to other genetic loci remains to be determined. This investigation explores the basic properties of location dependent transcriptional regulation by searching for both small and large chromosomal areas with correlated gene expressions.

The data we analyzed were collected by Bozdech et al. [5] who also considered the problem of spatial correlation. They reported finding a few regions of 2–7 genes each showing spatial correlation among the 14 linear chromosomes of *P. falciparum*. One limitation of their approach is that the search was based on correlated expression of adjacent genes independent of the physical distance between them. Thus, neighboring genes may be close or far apart. This approach might lead to a loss of power in detecting spatial correlation. Aburanti et al. [1] found correlations in *E. coli* by similar methods. In both investigations the main results were largely descriptive without a formal framework for statistical significance, especially with respect to the multiple testing aspects. One important observation made in this problem was by Balázsi et al. [2] who discuss the potential for spurious spatial correlation on the chromosome due to the spatial arrangement of probes on the microarrays themselves. In addition to suggesting detection methods for such artifacts, they also propose numerical methods to minimize this type of experimental bias. Kluger et al. [17] also report similar cautionary measures. Using signal processing methods, Jeong, Ahn, and Khodursky [16] report a thorough investigation of spatial chromosomal patterns in *E. coli* and provide convincing evidence of higher-order organization of transcription in bacteria.

Analytical Objective.   The overall objective of our work was to develop a visual and statistical framework to examine the correlation between gene expression and gene location. To this end, we develop methods to (1) explore general correlation patterns through a formal covariogram function, (2) assess regional correlation along each chromosome, and (3) assess and account for possible spatial trends throughout the array slide that may lead to spurious correlation between gene expression and gene location on the chromosome. Formal inference for significance is jointly accomplished with a permutation method and an adjustment for multiple testing via the false discovery rate method.

2.   METHODS

The data used in this investigation were collected by Bozdech et al. [5]. Briefly, a 48-hour time-course expression study was conducted using long-oligonucleotide arrays. At each time point the reference sample was a mixture of DNA from all 48 time points. For our own analysis we constructed an average profile for each gene by taking the pointwise average of all probes for a given gene.

2.1.   Data Pre-Processing

To perform our analysis it was necessary to create a data matrix that combined information from both gene expression and gene location. The normalized quality-control microarray data produced by Bozdech et al. [5] was combined with annotated nucleotide sequence information [13] to create a joint dataset. We achieved this by matching the gene identifiers from the gene expression dataset with the annotations from the sequencing centers contributing to the sequence data (version 2, October 3, 2002). Unique gene identifiers were found at plasmodb.org. The matching process resulted in the creation of a data matrix for 3495 unique genes. The data matrix was comprised of gene identifiers (rows) and gene information, including location, length, direction, and time-course expression. The start of a gene was defined as the end of the open reading frame closest to the 5′ end of its strand. For example, an open reading frame over base-pairs 100–200 would start at base-pair 100 if it were located on the Watson strand and at base-pair 200 if it were located on the Crick strand. For this investigation, the Watson strand is the reference strand, with all chromosomal locations listed with respect to the Watson strand.

To allow for the possibility of artifactual correlation between expression and chromosome location due to experimental design, we generated a time-course dataset that was adjusted for microarray probe location. We used a 2-dimensional nonparametric regression method (loess with $f = 0.05$, in the statistical package R [21]) to adjust the raw log-ratios (at the probe level) by regressing them on the (x,y)-coordinates of their array location. We then centered and scaled the adjusted ratios [24]. The resulting adjusted data correlated well with the quality control data of Bozdech et al. and the conclusions based on the two datasets were also similar. As such, the results reported here are limited to our adjusted data.

2.2.   Correlation Analysis

In this investigation we used Pearson correlation as a measure of distance between two expression profiles.

2.2.1.   Covariogram.   To examine the overall relationship between expression and chromosome location we used a covariogram, a correlation measure as a function of distance [11]. A covariogram function (γ) gives the correlation between genes x and y given that they are d_0 base-pairs apart:

$$\gamma(x, y; d_0) = r\left(x, y \mid \text{distance}(x, y) = d_0\right).$$

One typical assumption is that γ is homogeneous with respect to location; that is, we assume a constant correlation between genes x and y that are d_0 base-pairs apart, no matter where the genes may be located on the chromosome, the only important factor being that they are d_0 bp apart. We view distance symmetrically in the $3'$–$5'$ and $5'$–$3'$ directions, and ignore strand information. A consequence of the homogeneity assumption is that we average correlations across the entire chromosome, which are based on pairs of genes d_0 bp apart. A covariogram was created for each chromosome.

We also constructed covariograms giving correlations based on the physical location of genes on the microarray chip. This chip covariogram helps us interpret any observed correlations between expression and chromosome location. Having both types of covariograms can help us decide if the source of chromosomal correlation is due to biological phenomenon or is perhaps an

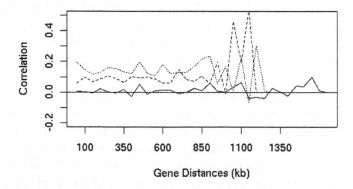

Figure 1. Covariogram by chromosomal distance for chromosome 6 (dotted), chromosome 4 (dashed) and chromosome 10 (solid).

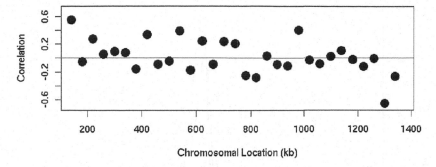

Figure 2. Correlation statistics calculated for chromosome 6 on a 40 kb interval.

artifact of chip design. For example, if the microarray is printed in gene order, then chromosomal correlations and printing errors, such as carryover, are likely confounded.

2.2.2.   Chromosomal Correlation Statistic.   It is important to directly account for the distance between adjacent genes when evaluating spatial correlation. To this end, we have partitioned each chromosome into non-overlapping contiguous intervals of resolutions 10, 20, 40, 60, 80 and 100 kb, and summarized pairwise correlations within each interval. Thus if two genes are adjacent neighbors 75 kb apart, neither one will influence measures of spatial correlation when small (e.g., 20 kb) regions are being explored. The notion of neighbor is therefore restricted to physical distance and not adjacency. Formally, we choose a partition size, p (bp), which for each chromosome yields n_p contiguous non-overlapping intervals. For each of the $i = 1, 2, \ldots, n_p$ intervals we calculate the average pairwise Pearson correlation, r_i, among the n_i genes in the given interval:

$$r_i = \frac{1}{k_i} \sum_{j=1}^{k_i} r_j, \qquad k_i = \binom{n_i}{2}.$$

For example, in an interval with 7 genes there are 21 pairwise correlations averaged to get r_i. This is repeated for every interval of each resolution on each chromosome. The results for chromosome 6 using a 40 kb partitioning are shown in Figure 2; significance is discussed below. To investigate possible bias due to starting the partition at the beginning of each chromosome we started at four different points. No obvious bias was observed; results with (Figure 4) and without (Figure 5) multiple starting locations are shown for chromosome 4.

2.2.3.   Permutation Test to Assign Significance.   To test the null hypothesis of no correlation between gene expression and gene location we de-

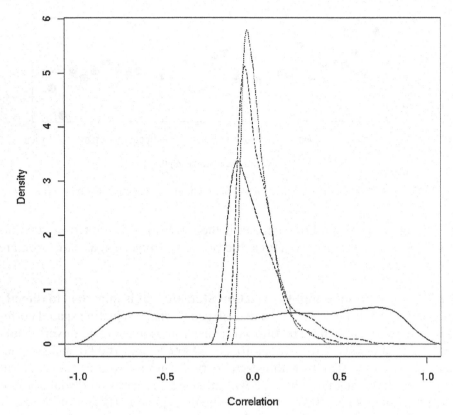

Figure 3. Null distribution for the statistic calculated from 2 (solid), 6 (long dash), 9 (short dash), and 12 (dotted) genes in different 40 kb intervals on chromosome 6.

veloped a permutation test [9,23]. For each chromosome and partition size (p) the gene orderings were kept fixed and the gene expression profiles were permuted to give observed values of r_i under the null hypothesis. The permutations are performed independently within each chromosome. Repeating this process $B = 1000$ times generated a null distribution for r_i, for each interval within each chromosome. To visualize the null distributions we used a kernel density estimate with a biweight kernel and the default bandwidth provided by R [21]. Examples for chromosome 6 are given in Figure 3. The observed statistic for a 9 gene interval based on a 40 kb partition was 0.49. Out of $B = 1000$ permutations a more extreme correlation occurred only two times, giving an approximate p-value of 0.002. Although we visualize the null distribution with smooth kernel density estimates, we directly compute p-values using the $B = 1000$ observed values of r_i: estimated p-value = $\#\{r_b \geqslant r_i\}/B$, where $b = 1, \ldots, B$.

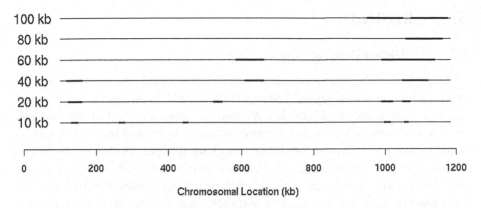

Figure 4. Significant intervals before correction for multiple testing on chromosome 4 at resolutions of 10, 20, 40, 60, 80 and 100 kb. Here the pointwise type I error level is 0.005. Note that at each resolution there were 4 equally spaced starting points resulting in overplotting of some intervals; compare with Figure 5, which has only one starting point for each resolution.

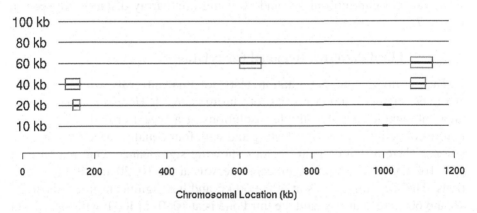

Figure 5. Significant intervals before (dark black line) and after (open boxes) correction for multiple testing on chromosome 4 at resolutions of 10, 20, 40, 60, 80 and 100 kb. Here the type I error level is 0.005 before multiple testing and after correcting there is an FDR of 0.05. In this partitioning scheme we have removed the overlapping intervals; there is one partitioning for each resolution which begins at 0 kb.

To correct for multiple testing we chose to control the false discovery rate (FDR) [3] for each partition (p) for each chromosome. Therefore, we account for multiple testing for each chromosome separately. This accommodates the independence of chromosomes with respect to correlation, which is supported by the covariograms. For each chromosome the algorithm orders the p-values, $p_{(i)}$ ($i = 1, \ldots, n_p$), and for a chromosome-wide level α a cutoff (k) is chosen, $k = \max\{i: \ p_{(i)} \leqslant \alpha i / n_p\}$, to determine significance at each interval. The k hypotheses $H^0_{(1)}, \ldots H^0_{(k)}$ corresponding to the k lowest p-values are rejected.

3.   RESULTS

3.1.   Global Patterns: Covariograms

To explore global trends of spatial correlation on each chromosome we constructed covariograms. Figure 1 shows three sample covariograms corresponding to chromosomes 6, 4, and 10. We find no apparent trend in chromosome 10, while chromosomes 4 and 6 show moderate but consistent correlation of approximately 0.2 throughout the distances up to three-fourths of their chromosome length. Covariograms for chromosomes 2, 4, 5, 9 and 10 also suggest correlation over large portions of their chromosomes. On balance, this exploratory data analysis indicates that there may be regions of spatial correlation.

To investigate spatial printing effects we also constructed covariograms based on microarray distance between the (x, y) coordinates of the probes for each gene. We found no evidence of correlation on the microarray. Covariograms of chromosomes 6 and 10 using microarray distance are seen in Figure 7.

3.2.   Local Patterns: Regional Correlation

The results of the permutation tests identify intervals with significant p-values. Figure 4 shows results for chromosome 4. Here we found several areas of significance at multiple resolutions at a type I error level of 0.005 before correcting for multiple testing and with four equally spaced starting positions. There are three main regions showing significance. One region is at 120–160 kb, which shows significant intervals at the 10, 20 and 40 kb resolutions. The second region is at 600–660 kb and has significant intervals at the 40 and 60 kb resolutions, and the third one is at 1080–1140 kb with significant intervals at all resolutions. After correcting for multiple testing and restricting to only one starting position, which yields a single partitioning of the chromosome, we arrive at the intervals in Figure 5. Boxed intervals are still significant after multiple testing. To further examine the data in these significant intervals, we plotted (Figure 6) the expression profiles corresponding to three different significant intervals. The left-hand panel shows four profiles from a region at 120–160 kb that was significant at resolution $p = 40$ kb. Below this panel is a diagram showing the physical array location of the probes corresponding to the time-course profiles. The middle panel shows six profiles from a region at 600–660 kb that was significant at resolution $p = 60$ kb. The right panel shows eight profiles from a region at 1080–1140 kb that was significant at resolution $p = 60$ kb. No obvious patterns of spatial correlation on the array chip were detected for two of these examples, however there may be spurious correlation at the region 120–160 kb due to printing artifacts. Other significant intervals

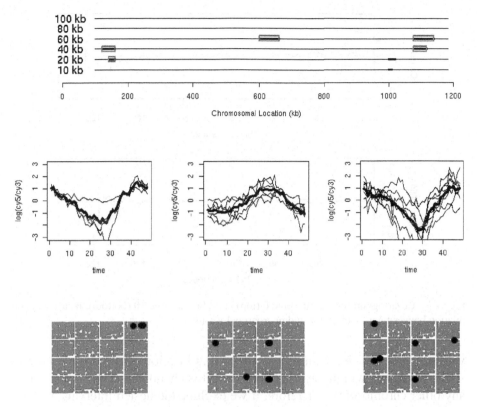

Figure 6. The results for chromosome 4 from Figure 5 are continued here. The left-hand panel shows four profiles from a region at 120–160 kb that was significant at resolution $p = 40$ kb. Below this panel is a diagram showing the physical array location of the probes corresponding to the time-course profiles. The middle panel shows six profiles from a region at 600–660 kb that was significant at resolution $p = 60$ kb. The right panel shows eight profiles from a region at 1080–1140 kb that was significant at resolution $p = 60$ kb.

showed (data not shown) the same "random" spatial arrangements on the array slide. This means that findings of spatial correlation on the chromosome are likely not confounded with printing artifacts, and thus point to biological explanations. Significant p-values for all chromosomes before and after multiple testing are available at ftp://ftp.stat.rice.edu/pub/blairc/CAMDA/.

These results were also corrected for multiple testing. We found 17 significant intervals with an FDR of 0.05. Five of these significant intervals occurred on chromosome 4, two each on chromosomes 5, 8, 9, 14, and one each on chromosomes 1, 10, 12 and 13. There were no significant intervals found on chromosomes 2, 3, 6, 7 and 11. The counting of significant intervals depends on the partition resolution. In Figure 5, for example, there are two significant intervals at ∼150 kb of chromosome 4, one at 20 kb partitions and another at 40 kb partitions. However, the 20 kb interval is nested within the 40 kb inter-

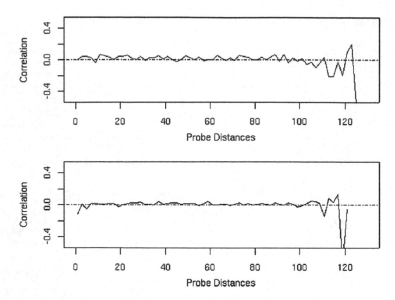

Figure 7. Covariogram for chromosome 6 (top) and chromosome 10 (bottom), measuring correlation between gene expression and microarray location.

val. Similarly, the significant 40 kb interval at location ~1100 kb of chromosome 4 is nested within the 60 kb interval. This phenomenon did not occur on any other chromosome. Therefore, if we exclude the nested intervals we have 15 significant intervals. In our discussion (below) of biological interpretation we refer to the 15 significant intervals.

3.3.   Interpretation of Results

A statistical finding does not guarantee a biological finding. It is therefore necessary to consider possible biological explanations. To this end, we examined ontology information for the significant regions. Gene Ontology (GO) is a formalized approach to organize information concerning gene products. The GO itself is a directed acyclic graph. Nodes in the graph are connected to each other hierarchically and represent terms to which individual gene products can be annotated. There are three main branches to the GO system: biological process, molecular function and cellular component. These main branches describe the main role of a gene product in the cell and downstream branches give more detailed characteristics. We mapped the plasmodium array onto the GO data structure using source information provided in a flat file (Pfa3D7_2002.10.03_Annotated_Annotation-v1.tab) from www.plasmodb.org. This gave a mapping from PFA gene IDs to GO nodes (terminal GO annotations). From this mapping we are able to recover full path information for all genes.

Table 1. Application of gene ontology information to significant intervals. Below are the chromosomal location (1–2), categories of GO similarities (3), number of genes annotated/total number of genes (A/T) in the interval (4) and description of similarities (5)

Chr	Location	GO similarity	A/T	Details
1	190–200 kb	Function	2/2	2 genes coding for transcriptional regulation
4	140–160 kb	Location, Function	2/3	2 genes annotated as cell surface proteins
4	1080–1140 kb	Location, Function	5/8	4 genes annotated as cell surface proteins
4	600–660 kb	Function	4/6	3 genes related to chromatin binding,
5	500–600 kb	Function	9/21	3 genes are GTP related
5	80–90 kb	Tandem Repeat	2/2	2 genes appear as tandem repeats of rhoptry-associated protein
8	300–400 kb	Function	8/18	6 genes associated with RNA processing
8	120–180 kb	Function	6/14	5 genes associated with protein synthesis
10	1400–1440 kb	Location, Function	3/6	3 genes annotated as cell surface proteins
12	1620–1680 kb	Function	3/6	3 genes associated with mitochondrial development
14	280–290 kb	Tandem Repeat	2/2	2 genes appear as tandem repeats of plasmepsin

Examination of the ontology information for the significant regions indicates that the proposed statistical methodology is able to detect possible biological relationships among the genes in the interval. In 11 of the 15 significant intervals we find biological similarities in putative function or in the cell location among the gene products (within each interval). In the remaining four significant intervals, three lacked ontology information and one showed no obvious similarities in cell location or function. Among the genes within each of the 11 significant intervals we found various biological similarities (Table 1), including those defined by possible tandem repeats, cell surface location of proteins, and similar function. The possible tandem repeats were observed on chromosomes 5 and 14, and clusters of surface proteins were seen on chromosomes 4 and 10. There were 6 intervals within which each contained genes annotated with similar function. These include a pair of transcriptional regulation genes on chromosome 1, chromatin binding genes on chromosome 4, small GTPase related genes on chromosome 5, RNA processing genes on chromosome 8, protein synthesis genes on chromosome 8 and mitochondrial development genes on chromosome 12.

To examine the sensitivity of our analysis to the choice of designed time points, we performed the identical analysis on only the even time points and on a random selection of 33 (70%) time points sampled without replacement. In the even (random 70%) time point analysis we found 16 (11) significant intervals with 11 (8) coinciding with the original 15 significant intervals based on all the time points. Most of the overlaps were on chromosomes 4 and 8. The original results do not appear to be an artifact due to time point design since we tend to detect a common pool of intervals in the three different analyses.

4.   CONCLUSIONS

Spatial correlation between gene expression profiles and chromosomal location may be defined in several ways. Considering adjacent pairwise correlations ignores inter-gene distance and thus may result in a loss of power to detect spatial correlation. Accounting for distance by restricting adjacent genes to be within a certain distance (bp) or through a formal covariogram function should provide more meaningful results and increase power. We have considered and compared the three approaches in the context of the *P. falciparum* time-course array study of Bozdech et al. [5]. Unlike previously reported findings we do find evidence of spatial correlation after accounting for inter-gene distance. Critical to the findings is a measure of statistical inference which we have implemented with a permutation approach.

Covariograms can be used as an exploratory tool to investigate correlation at small to large distances in one snapshot. Several chromosomes indicated short-range correlation as might be expected. However, we also found long-range moderate but consistent correlation ($r \sim 0.2$) in some chromosomes. This observation led us to examine larger window widths in our distance-based correlation investigations, resulting in several relatively long "blocks" of spatial correlation. This indicates that there may be some related function in these regions, or perhaps that there are silenced regions [7] along the chromosome. After correcting for multiple testing we still maintain several regions with strong evidence of spatial correlation, which we believe to be free of printing effects. Detecting spatial correlation due to biological function is the primary application of the methodology. However, an added benefit is the ability to detect possible errors in annotation as may occur when a single gene is accidentally annotated as multiple neighboring genes. In this case we would expect strong spatial correlation.

Considering more closely at each chromosome we found several areas throughout the genome that have significance at several resolutions of interval partitioning. At a nominal level of 0.005 we find several consistent results across resolution levels as indicated by the appearance of 'tornado' patterns (Figure 4, $\sim$1.1 Mb). Having annotation is crucial in assessing the biological significance of these findings. The Shaw lab generated a gene ontology

(GO) database for *P. falciparum* that allowed us to annotate our spatial correlation findings. Some of the most interesting results were gene similarities in function (e.g., transcriptional regulation, RNA processing, chromatin binding, mitochondrial development) and location of gene products on the cell surface. These types of result can help assess which regions are worth pursuing for further investigation. Although there are many annotations, many more of these potentially interesting intervals still lack annotation. Also interesting are intervals that have genes residing on the same DNA strand as they may provide clues to polycistronic regions, and several of these were also detected.

The notion that a chromosomal segment can have correlated gene expression has only recently been proposed for larger eukaryotic genomes [4,8,18,19, 22]. Bacteria and a few eukaryotes have operons in which multiple protein coding regions share a single upstream promoter from which a single polycistronic mRNA is initiated. However, the best characterized mechanism of expression control for eukaryotic genes involves locally acting regulatory DNA motifs called by different names such as upstream activating sequences [15] or enhancers [25]. Enhancers can act at some distance from the RNA polymerase binding site or promoter, but in almost all cases, enhancers regulate just one gene, typically by a looping mechanism that allows proteins bound at a distant enhancer to interact with the basic transcription apparatus at the promoter elements. The distance-correlated expression we find at several locations in the *Plasmodium* chromosomes cannot be due to a single enhancer, defined as described above. Instead, there must be a locally repeated regulatory motif [4], or a single regulatory motif that acts at a varying distance from the several local genes. One precedent for the latter model is the locus control region (LCR) that mediates coordinate regulation of the β-globin gene cluster in the mouse genome [6]. The LCR and similarly acting motifs are proposed to help organize dynamic domains of gene expression through localization of looped chromosomal segments in specific territories of the nucleus where gene expression is more active [18,19]. Whether similar mechanisms are present in *Plasmodium* remains to be determined.

This investigation provides evidence of chromosomal spatial correlation in gene expression in *P. falciparum*, and this correlation appears to be due to biological phenomena. There appear to be multiple levels of correlations, occurring at both small, mid and large scale. To the best of our knowledge ours is the first formal inferential method for statistically analyzing spatial correlation based on sequence and gene expression data.

ACKNOWLEDGEMENTS

This study was partially supported by NSF grant MCB 0091236 to M.C.G.; a training fellowship from the Keck Center for Computational and Structural

Biology of the Gulf Coast Consortia (National Library of Medicine Grant No. 5T15LM07093) to J.B.C.; NIH training grant T32CA096520 to J.N.-M. and R.G.

REFERENCES

[1] Aburatani, S., Sugaya, N., Murakami, H., Sato, M., and Horimoto, K., Statistical analysis of the relationship between gene expression and location, *Genome Informatics*, **14** (2003), 306–307.

[2] Balázsi, G., Kay, K.A., Barabási, A., and Oltvai, A.N., Spurious spatial periodicity of co-expression in microarray data due to printing design, *Nucleic Acids Research*, **31**(15) (2003), 4425–4433.

[3] Benjamini, Y. and Hochberg, Y., Controlling the false discovery rate: A practical and powerful approach to multiple testing, *Journal of the Royal Statistical Society B*, **57** (1995), 289–300.

[4] Boutanaev, A.M., Kalmykova, A.I., Shevelyov, Y.Y., and Nurminsky, D.I., Large clusters of co-expressed genes in the Drosophila genome, *Nature*, **420** (2002), 666–669.

[5] Bozdech, Z., Llinas, M., Pulliam, B.L., Wong, E.D., Zhu, J., and DeRisi, J.L., The transcriptome of the intraerythrocytic development cycle of *Plasmodium falciparum*, *PLoS Biology*, **1** (2003), 1–16.

[6] Bulger, M. and Groudine, M., Looping versus linking: Toward a model for long-distance gene activation, *Genes & Development*, **13** (1999), 2465–2477.

[7] Calderwood, M.S., Gannoun-Zaki, L., Wellems, T.E., and Deitsch, K.W., *Plasmodium falciparum var* genes are regulated by two Regions with separate promoters, one upstream of the coding region and a second within the intron, *Journal of Biological Chemistry*, **278**(36) (2003), 34125–34132.

[8] Caron, H., van Schaik, B., van der Mee, M., Baas, F., Riggins, G. et al., The human transcriptome map: Clustering of highly expressed genes in chromosomal domains, *Science*, **291** (2001), 1289–1292.

[9] Churchill, G.A. and Doerge, R.W., Empirical threshold values for quantitative trait mapping, *Genetics*, **138** (1994), 963–971.

[10] Cohen, B.A., Mitra, R.D., Hughes, J.D., and Church, G.M., A computational analysis of whole-genome expression data reveals chromosomal domains of gene expression, *Nature Genetics*, **26** (2000), 183–186.

[11] Cressie, N.A.C., *Statistics for Spatial Data*, J. Wiley, New York, 1993.

[12] Dever, T.E., Translation initiation: Adept at adapting, *Trends in Biochemical Sciences*, **10** (1999), 398–403.

[13] Gardner, M.J., et al., Genome sequence of the human malaria parasite *Plasmodium falciparum*, *Nature*, **419**(6906) (2002), 498–511.

[14] Grewal, S.I.S. and Moazed, D., Heterochromatin and epigenetic control of gene expression, *Science*, **301**(5634) (2003), 798–802.

[15] Guarente, L. and Ptashne, M., Fusion of *Escherichia coli lacZ* to the *cytochrome c* gene of *Saccharomyces cerevisiae*, *Proceedings of the National Academy of Science of the USA*, **78** (1981), 2199–2203.

[16] Jeong, K.S., Ahn, J., and Khodursky, A.B., Spatial patterns of transcriptional activity in the chromosome of *Escherichia coli*, *Genome Biology*, **5** (2004), R86.

[17] Kluger, Y., Yu, H., Qian, J., and Gerstein, M., Relationship between gene co-expression and probe localization on microarray slides, *BMC Genomics*, **4**(1) (2003), 49.

[18] Kosak, S.T., and Groudine, M., Form follows function: The genomic organization of cellular differentiation, *Genes & Development*, **18** (2004), 1371–1384.

[19] Kosak, S.T. and Groudine, M., Gene order and dynamic domains, *Science*, **306** (2004), 644–647.

[20] Neter, J., Kutner, M.H., Nachtsheim, C.J., and Wasserman, W., *Applied Linear Statistical Models*, The McGraw-Hill Companies, Inc., Boston, MA, 1996.

[21] R Development Core Team, *R: A Language and Environment for Statistical Computing*, R Foundation for Statistical Computing, Vienna, Austria, 2004.

[22] Spellman, P.T. and Rubin, G.M., Evidence for large domains of similarly expressed genes in the *Drosophila* genome, *Journal of Biology*, **1** (2002), 5.

[23] Wan, Y., Cohen, J., and Guerra, R., A permutation test for the robust sib pair linkage method, *Annals of Human Genetics*, **61** (1997), 79–87.

[24] Yang, Y.H., Dudoit, S., Luu, P., Lin, D.M., Peng, V. et al., Normalization for cDNA microarray data: A robust composite method addressing single and multiple slide systematic variation, *Nucleic Acids Research*, **30**(4) (2002), e15.

[25] Yaniv, M., Enhancing elements for activation of eukaryotic promoters, *Nature*, **297** (1982), 17.

Index